THE PROMETHEAN PURSUIT IN THE US-CHINA COMPETITION FOR GLOBAL TECHNOLOGICAL LEADERSHIP

THE PROMETHEAN PURSUIT IN THE US-CHINA COMPETITION FOR GLOBAL TECHNOLOGICAL LEADERSHIP

Khor Eng Lee, Aaron
& Bruno Khor

To order additional copies of this book, contact:
Xlibris
AU TFN: 1 800 844 927 (Toll Free inside Australia)
AU Local: (02) 8310 8187 (+61 2 8310 8187 from outside Australia)
www.Xlibris.com.au
Orders@Xlibris.com.au
840830

Contents

Foreword
By Teh Chor Aun

To quote from an in-depth and erudite article on history of technology in **BRITANNICA**:

"Civilization flourished continuously in China from about 2000 BCE, when the first of the historical dynasties emerged. From the beginning it was a civilization that valued technological skill in the form of hydraulic engineering for its survival depended on controlling the enriching but destructive floods of the Huang He (Yellow River). Other technologies appeared at a remarkably early date, including the casting of iron, the production of porcelain, and the manufacture of brass and paper...

"There were profound political changes in the 20th century related to technological capacity and leadership. It may be an exaggeration to regard the 20th century as "the American century," but the rise of the United States as a superstate was sufficiently rapid and dramatic to excuse the hyperbole... Technological leadership passed from Britain and the European nations to the United States in the course of these (two World) wars..." World War I (1914-18) and World War II (1939-45).

However, technological leadership is not a permanent property of any one nation, no matter how advanced and resourceful. Change is in the natural flow and order of things. Should the US continue to lose ground in its discernible decline, China is poised to reclaim global technological leadership in a matter of time.

In Greek mythology, the Titan Prometheus in his immense love for humanity, stole the sacred fire from the gods to give it as a gift to the humans for their immeasurable benefit as well as a lasting boon and hope for them to struggle for a better future.

It was the first technology theft recorded in literature.

Fire symbolizes knowledge as well as the fountain-head of technology.

"I've always thought of AI (artificial intelligence) as the most profound technology humanity is working on -- more profound than fire or electricity or anything that we've done in the past," Alphabet and Google CEO Sundar Pichai has said.

Google's launch of its AI chatbot Bard 21 March 2023, following the historic debut of OpenAI's ChatGPT 30 November 2022, has sparked the new stage of the global high-tech race from generative AI to AGI (artificial general intelligence) -- what AI mega-investor Ian Hogarth has called "God-like AI".

To quote Yana Khare (April 20, 2023 **Analytics Vidhya**): "Just as humanity has harnessed the power of fire and electricity for the greater good, we must also rise to the challenge of responsibly developing and deploying AI for the benefit of all."

On 11 April 2023, China's cyberspace authority Cyberspace Administration of China (CAC) issued a content regulation draft for public comments -- to "promote the healthy development and standardized application of generative AI."

And China officially supports the independent innovation, popularization and application as well as international cooperation in basic technologies such as AI algorithms and frameworks.

China also encourages the priority use of safe and reliable software, tools, computing, and data resources. (Shen Weiduo and iong Xinyi Apr 11, 2023 **GLOBAL TIMES**)

On 18 April 2023, China Mobile Communications Association (CMCA) and other industry players including telecom giants like China Mobile and China Unicom jointly established the GPT industry Alliance in Beijing to nurture and grow a more "inclusive, secure, and self-developed universal artificial intelligence (AI) in China.

And, the alliance will build "a solid bridge between the government, academia and industry, giving birth to China's independent universal AI." (**Shanghai Securities News/Global Times** Apr 18, 2023)

Major technology companies like Alibaba, Baidu and Tencent have reportedly announced plans in releasing ChatGPT-like chatbots and in integrating their services into consumer applications.

"We have to accept the <u>challenge of AI</u>, gain the initiative of the digital intelligence era, and make China the world's <u>innovation center</u> in the global 'intelligence' game," said Ni Jianzhong, executive president of the CMCA at the inauguration of the GPT Industry Alliance.

21.04.2023 18:03 p5/105 23025 words

On 07 May 2021, Liu Qingfeng, chairman of iFlytek, introduced its generative AL large language model SparkDesk through a live demonstration in Hefei, East China's Anhui province. (**Global Times** May 07, 2023)

"We aim to exceed ChatGPT capabilities in Chinese language by October 24, and will achieve a similar level to it in English," Liu said. "The impact of this generative AI technology is no less important than that of the birth of the PC (computer) or the internet…"

10.05.2023 19:10

Today, the true Promethean pursuit is in the single-minded quest for new technological knowledge and skills, to make fresh discoveries and to innovate and turn out novel technologies for the benefit of humanity, but certainly not to divide human races, countries and nations, nor to drive them into counterproductive collusion cum collision and possible disruptive/destructive confrontation and conflict.

The traditional Chinese way is the civilized practice of wisdom which can lead to win-win collaboration and cooperation for mutual benefit, for the common weal of humanity and a shared future for the entire community of the world's nations.

Let's pray for common sense, good will, and durable peace. May humanity prevail. Amen.

10.11.2022 11:52 20;41
05.02.2023 15;53
22.04.2023 19:16
11.05.2023 21:12

1

The US-China Strategic Competition

Dr Miles Maochun Yu, a key advisor to US Secretary of State Mike Pompeo in the Trump administration, said to Bill Gertz in their Q & A (**The Washington Times** Monday June 15, 2020): "At the outset of this (Trump) administration, the White House revamped our strategic outlook and the president issued a landmark and far-reaching National Security Strategy in December 2017. Subsequently, the Pentagon issued the companion document known as the National Defense Strategy.

"Both documents ushered in an age of Great Power Competition in which China no longer will be treated by the U.S. as merely a "card" to be played in order to reach other strategic goals. Instead, China is at the top of our national security agenda, as there is no bigger threat than China and no other more important strategic goal than stemming China's threat to the world…" 03.05.2023 20:59

In Commentary in **WAR ON THE ROCKS** August 31, 2022, Dr Cheung Tai Ming and Dr Thomas G. Mahnken have written. (having co-edited July 2018 publication "THE GATHERING PACIFIC STORM Emerging U.S.-China Strategic Competition in Defense Technological and Industrial Development):

"… Although both the Bush and Obama administrations expressed concerns about the growth of Chinese military power, it was not until the Trump administration that documents such as the National Security Strategy (December 2017) and National Defense Strategy (January 2018) spoke openly about the challenge posed by China and made great power competition the foremost priority. The Biden administration views China as "our most consequential strategic competitor and the pacing challenge" in its defense planning. Although today there is general consensus on the need to counter its aim to become a high-technology superpower, action has lagged rhetoric…"

There was a perceived lack of action before Biden unleashed the storm of US legislative forces to out-Trump and outdo his tempestuous predecessor in a self=ordained mission to ravage the economic and technological domains of his formidable rival in the East.

"Today, we are emerging from a period of strategic atrophy, aware that our competitive military advantage has been eroding," the 2018 National Defense Strategy has stated. "The <u>reemergence of long-term strategic competition</u>, rapid dispersion of technologies, and new concepts of warfare and competition that span the entire spectrum of conflict require a Joint Force structured to match this reality…"

p2 267 words 27.10.2022 17:57

The highly reputable RAND Corporation recently carried out a lot of insightful research on the US-China strategic competition.

"For the first time since contending with the Soviet Union in the Cold War, the United States faces the prospect of a long-term competition with a near-peer great power: the People's Republic of China (PRC). China's economy has become (since 2010) the second-largest in the world, and its companies compete with U.S. counterparts for markets and resources. The People's Liberation Army (PLA) has become the "pacing threat" to U.S. military operations in Asia, and China's diplomatic influence rivals that of the United States in many parts of the world.

"The narrowing gap in national strength has coincided with an intensification of bilateral disputes over trade, technology transfer, cyber espionage, human rights, and other issues. Even the shared threat of the coronavirus disease 2019 (COVID-19) has proved an occasion for the two sides to trade accusations and compete for influence. Tensions have grown around smoldering hot-spot issues, such as Taiwan and the East and South China Seas," Timothy R. Heath, a senior international and defense researcher at the RAND Corporation, has reported on "U.S. Strategic Competition with China" in A RAND Research Primer June 2021.

"This rapid unraveling of the U.S.-China relationship -- which had been widely viewed as stable and mutually profitable despite its long-standing disputes -- has unsettled global politics. Although both capitals appear committed to peacefully resolving their differences, the intensifying acrimony and distrust have raised fears among many observers that the two countries could be headed toward confrontation…"

Heath has reported that RAND Corporation research over the past few years "sheds light on many aspects of the enormously complex and important U.S.-China strategic competition". With more than 60 RAND reports, primarily from the past five years to mid-2020, the RAND primer covers various issues from China's strategic goals and priorities, its policies and measures, to U.S. strategic interests & initiatives.

Main findings include the fact that the US has the upper hand in comprehensive national power (CNP) but China is narrowing the gap, the perilous erosion of the U.S. security position in Asia as a result of the PLA's advances in modernization, and the intensifying struggle through various means and measures other than war. 639 words 28.10.2022 11:27 m

On the high stakes and significance of "the most important competition" of the US with China, Heath has succinctly related the findings of various RAND studies as follows:

As one 2019 study by James Dobbins, Howard J. Shatz, and Ali Wyne put it, "China is a peer competitor that wants to shape an international order that it can aspire to dominate." Noting the possibility that Beijing might shape an international order to its economic benefit and to the detriment of U.S. economic prospects, the report described China as posing a "less immediate threat, but a much greater long-term threat…"

A 2018 RAND report by Michael J. Mazarr, Timothy R. Heath, and Astrid Stuth Cavallos similarly examined the implications for global peace from the deepening U.S.-China competition for influence within the international order. As that report noted, "Whether a growing competition for influence and leadership with the United States in shaping the terms of the international order escalates into dynamics that become destructive of that order remains to be determined."

Other studies have likewise highlighted the significance of the U.S.-China competition for shaping the evolution of the international system. A 2019 Rand study by Michael J. Mazarr and his colleagues characterized the current era as one featuring "an overarching competition with China, with secondary, largely regional contestations with other actors, including Russia." The study called the U.S.-China contest "decisive" for the overall character of international competition.

RAND researchers have paid particular attention to China's economic prowess and its willingness to exert diplomatic influence to strengthen its position. The 2019 report by Dobbins concluded as follows: "It is geoeconomics, rather than geopolitics, in which the contest for world leadership will play out." Specifically, "The principal Chinese challenge is not that it will impose authoritarian governments on its trading partners but that, over time, it will skew global standards for trade and investment in its favor to the disadvantage of its competitors. A 2020 study by Shatz similarly noted China's willingness to use economic tools to compete in the security and geopolitical domains.

With respect to technology, RAND researchers have emphasized the importance of leadership in the development and manufacturing of advanced technologies. Technological leadership could not only enable rapid economic growth but also allow a country's military to gain an edge on the battlefield.

RAND researchers have begun to examine particular technological sectors to investigate how much of a threat <u>China's pursuit of technological advantage</u> poses to U.S. industry.

For example, a 2017 study led by Chad J.R. Ohlandt judged that China's unambiguous policy drives "a whole-of-government effort to develop a globally competitive aviation industry." The study detected "few technology-transfer concerns" for the sector, noting that Chinese investments have been limited to "companies with technologies not particularly relevant to commercial or military aircraft." Nonetheless, the study recommended that export controls remain in place, given China's "aggressive aviation industrial policies" and clear intent to compete in the sector.

In conclusion, 6 areas for further research have been spotted, including (4) technological aspects of the U.S.China competition: The importance of technology in the competition has also gained considerable attention in RAND research.

The <u>competition for technological leadership</u> not only offers the potential prizes of economic gain and corporate profits but also carries strong implications for <u>military modernization.</u> Advanced technologies could dramatically reshape how wars are fought and confer a significant advantage on the side that masters the technologies.

Insights into how the U.S. government can effectively manage the technological aspect of the competition could prove an important determinant in the outcome of the U.S.-China strategic competition. 6p 1241 words 28.10.2022 13:32

In a synopsis of THE GATHERING PACIFIC STORM July 2018: US-China military technological competition lies at the heart of the growing strategic contest between the United States and China. This is largely because this technological rivalry straddles the geostrategic and geo-economic domains covering drivers ranging from industrial policy and foreign direct investment (FDI) to weapons development programs and threat assessments. Examining the nature of the US-China defense technological competition requires a more comprehensive and nuanced understanding of the complex military, economic, innovation, and other drivers at play. Moreover, this technological race is still in the early stages of development and can be expected to grow larger, more complex, and more intense…

"The Chinese government is fighting a <u>generational</u> fight to surpass our country in economic and technological leadership," FBI Director Christopher Wray said in opening **remarks at the China Initiative Conference organised by the Center for Strategic & International Studies (CSIS), Feb 6, 2020.**

"To surpass America, they (the Chinese) need to make leaps in cutting-edge technologies…:"

In a paper published on 24 June 2020 "Technology, power, and uncontrolled great power strategic competition between China and the United States" (Springer China International Strategy Review 2), Xiangning Wu of University of Macao has written: "… There is no need to argue that great power competition has returned to the global centre stage. The new round of competition is developing with unprecedented uncertainties.

"The fierce competition between China and the U.S. has already expanded from trade (Trump-triggered March 2018) to the protection of cutting-edge technologies, regional strategies, and two development models supported by different values. Geo-economics is the primary arena of great power competition, while geopolitics and technology are increasingly intertwined.

"Moreover, the two countries' divergence in social values and political systems is becoming more intense. As a result, technology is now largely politicized (and weaponized) and has become a

more prominent element of great power rivalry and politicians are racking their brains to assess the risks of and exaggerate the severity of conspiracy in high-tech cooperation with China...

"Because of the critical role of technology, in comprehensive economic development and national military power, great powers, particularly China and the U.S., will inevitably race to develop technology to pursue future global eminence.

"The political leaders of both China and the U.S. profoundly recognize that <u>technological innovation</u> is a potent source of national power. Both leaders realize that the importance of <u>technological leadership</u> cannot be overstated..."

At the 2021 Aspen Security Forum in Washington, D.C., 4 November 2021, Michael Brown, Director of Defense Innovation Unit (DIU), declared: "We need <u>technological advantage to prevail in this strategic competition</u> with China.

"For the military, that means we've got to modernize faster. We've got to use more commercial technology..." (U.S. Department of Defense Terri Moon Cronk Nov. 5, 2021)

"Beijing views the United States as its strategic competitor and a fundamental threat to its goal of achieving national rejuvenation by becoming a science and technology (S&T) leader.

"China's motivation to substitute foreign technology is thus not dependent on its perception of the dependability of US firms. It will seek to eliminate dependencies on US technology regardless of how restrictive US export controls may be," Gregory Graff, an analyst for the Department of Defense, has written in his report on COUNTERING CHINA'S MILITARY MODERNIZATION THROUGH TECHNOLOGY CONTROLS. Graff focuses on China's military and strategy.

His report concludes: "The United States and China are locked in strategic competition for composite national power (CNP), for which cutting-edge technology is a key enabler across economic and military domains. In response to China's growing technology levels and its whole-of-nation effort to absorb US technology, the US has pursued significant controls on China's commercial access to US technologies…"

(American Enterprise Institute October 2022):…

"… The PRC (People's Republic of China) has expanded and modernized nearly every aspect of the PLA (People's Liberation Army), with a focus on offsetting U.S. military advantages. The PRC is therefore the pacing challenge," the US Department of Defense (DoD) has declared on strategic competition with China. (2022 National Defense Strategy (NDS), October 27, 2022)

"In addition to expanding its conventional forces, the PLA is rapidly advancing and integrating its space, counterspace, cyber, electronic, and informational warfare capabilities to support its holistic approach to joint warfare. The PLA seeks to target the ability of the Joint Force to project power to defend vital U.S. interests and aid our Allies in a crisis or conflict.

"The PRC is also expanding the PLA's global footprint and working to establish a more robust overseas and basing infrastructure to allow it to project military power at greater distances. In parallel, the PRC is accelerating the modernization and expansion of its nuclear capabilities…"

In the 2022 Nuclear Posture Review (NPR), DoD has amplified: "The People's Republic of China (PRC) is the overall challenge for U.S. defense planning and a growing factor in evaluating our nuclear deterrent. The PRC has embarked on an ambitious expansion, modernization, and diversification of its nuclear forces and established a nascent <u>nuclear triad</u>. The PRC likely intends to possess at least 1,000 deliverable warheads by the end of the decade (2030)…" (According to Hans M. Kristensen and Matt Korda in <u>Bulletin of the Atomic Scientists</u> January 16, 2023, the US strategic force has deployed 3,508 warheads.)

To continue with the 2022 Nuclear Posture Review (NPR): "While the end state resulting from the PRC's specific choices with respect to its nuclear forces and strategy is uncertain, the trajectory of these efforts points to a large, diverse nuclear arsenal with a high degree of survivability, reliability, and effectiveness…"

To quote from the Doomsday Clock Statement (90 seconds to midnight/Armageddon) by <u>Bulletin of the Atomic Scientists</u>, January 24, 2023: "… The US Defense Department claims Beijing may increase its arsenal five-fold by 2035 (t0 over 1,500 warheads) and could soon rival the nuclear capabilities of the United States and Russia, with unpredictable consequences for stability…

"The United States, Russia, and China are now pursuing full-fledged nuclear wepons modernization programs, setting the table for a dangerous new "third nuclear age" of competition…"

The Pentagon has its message in the 2022 NPR script: "We will maintain a flexible deterrence strategy and force structure that continues to clearly convey to the PRC that the United States will not be deterred from defending our Allies and partners, or coerced into terminating a conflict on unacceptable terms…"

P87 18828 words 03.02.2023 20:55

In the 2022 NATIONAL SECURITY STRATEGY, President Joe Biden has written October 12, 2022:

"… The 2022 National Security Strategy outlines how my Administration will seize <u>this decisive decade</u> to advance America's vital interests, position the United States to outmaneuver our geopolitical competitors (principally to out-compete China and to constrain Russia), tackle shared challenges (e.g. climate change), and set our world firmly on a path toward a brighter and more hopeful tomorrow…

"… United States has everything we need to win the competition for the 21st century. We emerge stronger from every crisis. There is nothing beyond our capacity…:"

(NATIONAL SECURITY STRATEGY October 2022)

P10 1974 words 28.10.2022 18:46 29.10.2022 13:33

And to quote LOWY INSTITUTE ASIA POWER INDEX 2022 (p. 8): "The portrait that emerges is this: China's overall power still lags the United States but is not far behind. According to theorists, a power transition is triggered when a rising power's overall strength approaches 80% of that of the established power. The Asia Power Index showed in its inaugural edition that China had convincingly breached this threshold in 2018. Washington is unlikely ever to re-establish a decisive lead. The age of uncontested U.S. primacy in Asia is over…"

To further quote from the Australian think tank: "On current trends, China is now less likely to pull ahead of its rival in comprehensive power by the end of the decade (by 2030). Even if it does in future decades, it appears highly unlikely China will ever be as dominant as the United States once was.."

21.04.2023 18:52 p19/106 23175 words

On the opening day of the CPC's 20th National Congress 16 October 2022, **CCTV (China Central Television)** reported on President Xi Jinping's pledge to boost China's strength in strategic science and technology (S&T).

Xi's pledge is that the Chinese nation will prevail in its struggle to develop strategically important tech, underscoring Beijing's concern over the US campaign to separate it from cutting-edge chip capabilities.

05 November 2022 18:16
05.02.2023 16:34
22.04.2023 20:25
11.05.2023 21:27

2

US-China Competition for Technological Leadership

"… The United States and China are in a growing competition, perhaps verging on conflict, but it is not a nineteenth century competition between empires for control of territory and resources. Unlike great power competition in previous centuries, the focal point is not military strength or territorial expansion. This conflict is over <u>control of the modern levers of power</u> -- <u>global rules and institutions, standards, trade, and technology</u>," James Andrew Lewis, senior vice-president at the Center for Strategic & International Studies (CSIS), has written November 30, 2018.

"The ability to create new technologies, particularly digital technologies (given their importance for politics, security, and economic growth) have become key factors in the U.S.-China relationship, which is marked by close commercial cooperation and deep governmental distrust. This disparity creates unavoidable tensions…"

This article by Lewis on "Technological Competition and China" offers other relevant and salient insights:

The link between technology, innovation, national security, and international power is now widely recognized.

A country's ability to <u>innovate</u> and produce <u>advanced technologies</u> provides economic strength, military power, and an intangible benefit of <u>perceived leadership</u>.

Both China and the United States have advantages and disadvantages in this contest, and while it is usual to focus on quantitative aspects -- such as the number of engineers or patents and spending on research and development (R&D) -- these are not the key determinants of technological competition between states.

This competition is a contest of ideas on governance for investment, innovation, and the internet. The internet and global connectivity not only reshape the environment for competition but also create political and market forces that both nations find difficult to control.

Chinese leaders realize that technological leadership is one of the accoutrements of power, something first captured in China by the idea of "two bombs and one satellite" (in the early Mao era), which showed the Soviets and the Americans that China was a peer in strategic technology and did not need them.

The focus now is on surpassing the United States in a broad range of technologies while again asserting that China does not need US' help to do this (self-reliance and independence!)

Competition over technological "leadership" has become central -- which is not the case with other U.S. opponents, such as Russia or Iran.

Li Zheng of the China Institute of Contemporary International Relations has been quoted in saying: "The United States views technology as the 'last barrier' to constrain China's challenge." According to Li,, U.S. actions have "risen to the strategic level", seeking to "systematically and comprehensively curb the rapid rise of China's technology industry". (WITA Washington International Trade Association Adam Segal 06/01/2019)

"Chinese analysts and policy makers have interpreted U.S. efforts to prevent the flow (of) critical technologies through limits on investment, blocks on the operations of Huawei and other Chinese telecom companies in the U.S. and other markets, and new export control laws, as part of a strategy of containment designed to slow China's rise as a science and technology (S&T) power. In response, a newly emerging strategy consists of: a doubling down on indigenous innovation and developing "core technologies"; protection of supply chains; diversification of access to foreign technology; diplomatic efforts that stress the shared benefits of Chinese technology development; and continued cyber-enabled theft of intellectual property (IP)," Adam Segal, Ira A. Lipman chair in emerging technologies and national security, has written (06/01/2019 WITA). Segal is director of the Digital and Cyberspace Policy Program at the Council on Foreign Relations.

"Even though both sides are likely to lose the efficiencies that came from the globalization of innovation, such a strategy may also energize American and Chinese policy makers to mobilize even greater resources for scientific competition," Segal has suggested.

"Washington, anxious about China's rising technological capabilities and its program of military-civil fusion (MCF) has limited Chinese investment in U.S. technology sectors, blocked Chinese telecommunications companies from doing business in the United States and other markets, and tightened controls on the sale of technologies..."

According to most analysts, US "containment" policies are not a response to Beijing's industrial policies or IP theft, but rather they stem from a <u>decline in U.S. power and prestige</u> and a "panic" about China's rise in technologies, such as 5G and AI.

"Even if the US and China agree on a deal that would end their on-again, off-again trade war, the economic and trade relationship between these two countries will be fraught for years to come. This is because the constant dispute is not so much about trade, but rather about larger structural issues. The US and China are locked in <u>a race for economic and technological dominance</u> in the long-term. Resolving this new rivalry will require both sides to find a mutually acceptable middle ground," Marianne Schneider-Petsinger has written A research fellow in the US and the American Programme at Chatham House (an independent world-leading policy institute in Kent, England), she has written November 2019 on <u>The Race for Global Technological Leadership</u>.

"Many of the US-China tensions in the area of technology transfer, IP and innovation arise because of American concerns over China's ambition to become a global leader in a wide range of technologies. In particular, the industrial policy <u>'Made in China 2025'</u> -- which is aimed at expanding the <u>high-tech sector</u> in such fields as aerospace, robotics, and information and communications technology (ICT) -- is seen as a threat to US technological leadership. The Trump administration has described such Chinese policies as 'economic aggression' (threatening the technologies and intellectual property (IP) of the US) -- White House Office of Trade and Manufacturing Policy, June 2018).

"Another reason for the Trump administration's increasingly confrontational approach to China is that many of the next generation technologies have both civilian and military applications. Thus, US apprehensions go beyond purely commercial issues…

"The underlying driver of the ongoing US-China trade war is a race for global technological dominance…"

In The Heritage Foundation COMMENTARY Jan 28, 2020,

James Jay Carafano, Vice President, Kathryn and Shelby Cullom Davis Institute, and a leading expert on national security and foreign challenges, has commented on "Technology and Great Power Competition":

"… It is a competition without precedent. One aspect of this new world disorder -- <u>advanced technological competition</u> -- is particularly noteworthy. It will be infused throughout the struggle, and if Washington wants to win, America wil have to do better…

"Competing effectively means it's as important, if not more important, to sustain American strengths and competitive advantages…"

In July 2020 Eric Schmidt and Jared Cohen formed the China Study Group (of 15 persons) to advance policies "that position the United States to out-compete China without inviting escalatory cycles of confrontation, retaliation, or unintended conflict. While competition is the dominant frame, also essential is considering when cooperation, collaboration, and exchange with China is in our interest, as severing ties and closing off the United States to the ideas, people, technologies, and supply chains necessary to compete effectively will undermine U.S. innovation..."

The China Study Group (CSG) released its 35-page mission report in Fall 2020: "Asymmetric Competition: A Strategy for China & Technology Actionable Insights for American Leadership".

"America's technological leadership is fundamental to its security, prosperity, and democratic way of life. But this vital advantage is now at risk, with China surging to overtake the United States in critical areas. Urgent policy solutions are needed to renew American competitiveness and sustain critical U.S. technological advantages," CSG has reported.

"In order to lead, the United States will need to maximize its competitive advantage in key strategic technologies (from gene editing and multiplexing in biotechnology to biochips and other next generation chips) in ways that overcome China's advantages, which include greater scale, hyper-integrated platforms, and faster product loop...

"Platform dominance is a crucial aspect of competition with China.

"We need to upgrade our intelligence capabilities to dominate the forecasting space. If we cannot forecast when (& where) technology is going we cannot stay competitive..."

Matt Sheehan, a fellow at the Carnegie Endowment for International Peace, commented April 12, 2022: "... Competing with China in technology remains one of the great challenges facing America today, and my research is devoted to figuring out how to do that effectively..."

Sheehan's research focuses on global technology issues, with a specialization in China's AI ecosystem.

"When it comes to competing with China in technology, the greatest threat to U.S. leadership is not TikTok (a slingshot directed at Mark Zuckerberg's Meta). It's the deep political dysfunction and wrenching social divisions at home, some of which are accelerated by the worst excesses of big tech," Sheehan has boldly and insightfully pointed out.

"As these social, political, and economic divisions have been wrenched open, the country has teetered dangerously close to outright political collapse…"

Sheehan has also cogently pointed out that "fear of Chinese technology is counterproductive and threatens to do even greater harm to U.S. technological competitiveness". In brief, fear of Chinese competition won't preserve US tech leadership.

On 12 September 2022 the newly-established Special Competitive Studies Project (SCSP) issued a press statement on the publication of its first report "Mid-Decade Challenges to National Competitiveness". The SCSP is a private outfit founded late 2021 and led by former Google CEO Eric Schmidt and former deputy defense secretary Bob Work.

According to report, US is in a fierce technology competition with China that is shaping the future of geopolitics and the contest between democracies and autocracies.

"The PRC (People's Republic of China) is the United States' chief ideological opponent, largest economic competitor, most capable technology peer, and most threatening military rival. Technology is central to all parts of the competition," SCSP Board of Advisor Nadia Schadlow said.

The stark message is: the US could lose the competition if dramatic action is not taken across a broad range of public policy arenas to invest in U.S. technology advantages, strengthen the techno-industrial base, and deploy disruptive technologies democratically and responsibly.

"We cannot keep playing catch-up like we have on 5G and microelectronics supply chains," SCSP CEO Ylli Bajraktari argued. "The United States needs to organize, make strategic tech bets, help resource technology sectors and applications, and adapt our national security tools".

Said SCSP Board of Advisor Mae Thornberry, "Our success hinges on Americans recognizing the urgency and stakes of the competition…"

According to Bob Work, the US must win on three technology "battlegrounds" of microelectronics, 5G wireless technology, and AI. (Jaspreet Gill September 13, 2022 **BREAKING DEFENSE**)

Work has described the technological competition with China as "going to be the defining feature of global politics for the rest of our lives."

Work has also stressed: "And it is going to determine who is the greatest economic power in the 21st century. It's going to determine who is the greatest military power. It is a competition that we simply must win…"

In a prepared statement for hearing before the Committee on Armed Services United States Senate 117th Congress February 23, 2021 (**congress.gov**), Dr Eric E. Schmidt, Co-Founder, Schmidt Future, submitted: "… China is pursuing technological leadership through strategic investments in a wide range of critical technology areas (from AI and 5G to quantum computing and synthetic biotechnology), including through the Made in China 2025 initiative…"

Schmidt has subsequently described China as "an autocratic competitor that is run by technocrats, that is very capable of inventing a new future…" (**REUTERS** 4 April 2023)

06.04.2023 03:52

`Megan Lamberth and Martin Rasser of the Center for a New American Society (CNAS) have authored the CNAS Report "Taking the Helm: A National Technology Strategy to Meet the China Challenge".

"... China's leaders have made scientific and technological (S&T) leadership the focus area in its drive to become the world's economic dynamo, the power center of a new geopolitical order, and a global military leader," Lamberth and Rasser have written in **THE HILL** Opinion 09/07/21.

"How U.S. leaders act in response will determine whether America maintains its status as the world's preeminent scientific and technological (S&T) power, with all the advantages that confers, or descends on a slow glide path to mediocrity. Both outcomes are plausible, Status quo policies make the latter most likely…"

P16 3240 words 29.10.2022 17:37 p19 3968 words 30.10.2022 11:21
10.11.2022 10:42 05.02.2023 17:53 22.04.2023 21:07

POSTSCRIPT "The Chimera of Technological Superiority"
By Kelly A. Grieco and Robert A. Manning, March 21, 2023

"... No other nation places as much faith in technology as does the United States. The spirit of invention and discovery has powered America's economic and geopolitical strength, and it is deeply embedded in U.S. military culture. But technology is not strategy.

"Some will undoubtedly argue that the Pentagon should strive to maintain technological superiority, if only because China is doing the same in pursuit of its own military-technological edge. Nonetheless, general-purpose technologies like AI, which have a range of applications across industries, are not like a tank or fighter jet. These are "enabling technologies" akin to electricity or the steam engine, and any military advantage derived from them will be determined by how military organizations use them, rather (than) by the access to those technologies alone," Kelly A. Grieco and Robert A. Manning have persuasively written, arguing that the pursuit of military dominance through technological superiority, amid rapidly diffusing dual-use technologies, is based on flawed and unproven assumptions (Grand Strategy Red Cell Project, The Henry L. Stimson Center).

"The United States ought to remain competitive and invest in emerging technologies, as the administration's CHIPS ($52.7 billion CHIPS and Science Act of August 2022) and IRA ($369 billion Inflation Reduction Act of August 2022) legislation seek to do, but the Pentagon wants to chase after the next "shiny object," as if technology itself is strategy rather than a tool of national power. This technological fetishism only obscures the real challenge: shaping a strategy for a world in transition at a time of unprecedented technological change."

P29/106 232szzzzzzzz79 words 22.03.2023 01:56 22.04.2023 21:25 24.04.2023 00:15
12.05.2023 02:55

3

US-China Techno Competition: A Progress Report

"… Today, China's rapid rise to challenge U.S. dominance of technology's commanding heights has captured America's attention. The rivalry in technology is what the Director of the Central Intelligence Agency (CIA), Bill Burns, spotlights (October 2021) as the main arena for competition and rivalry with China."

"It (China) has displaced the U.S. as the world's top high-tech manufacturer, producing 250 million computers, 25 million automobiles, and 1.5 billion smartphones in 2020.

"Beyond becoming a manufacturing powerhouse, China has become a serious competitor in the foundational technologies of the 21st century: artificial intelligence (AI), 5G, quantum information science (QIS), semiconductors, biotechnology, and green energy. In some races, it has already become No. 1. In others, on current trajectories, it will overtake the U.S. within the next decade (by 2030)," the Harvard Kennedy School BELFER CENTER has reported in its paper The Great Tech Rivalry: China vs the U.S., Dec. 07, 2021. The report was authored by Graham Allison, Douglas Dillon Professor of Government at Harvard Kennedy School, with Kevin Klyman, Research Assistant at Belfer Center, and two former Research Assistants at Belfer Center -- Karina Barbesino and Hugo Yen.

To quote: "To take stock of the state of the technology race, this report examines the progress made by the U.S. and China in each key technology over the past 20 years.

"To begin with our <u>bottom lines</u> up front:

* In the advanced technology likely to have the greatest impact on economics and security in the decade ahead -- <u>AI</u> -- China is now a "full-spectrum peer competitor" in the words of Eric Schmidt (former Google CEO).

* In <u>5G</u>, according to the Pentagon's Defense Innovation Board (DIB), "China is on track to repeat in 5G what happened with the United States in 4G. Despite advantages in 5G standards and chip design, America's 5G infrastructure rollout is years behind China's, giving China a <u>first-mover advantage</u> in developing the 5G era platforms.

* In quantum information science (QIS), America has long been viewed as the leader, but China's national push presents a clear challenge. China has already surpassed the U.S. in quantum communication and has rapidly narrowed America's lead in quantum computing. (In 2017 the U.S.-China Economic and Security Review Commission reported that "China has closed the technological gap with the United States in quantum information science (QIS) -- a sector the United States has long dominated." Beijing has since named QIS as a top tech priority second only to AI in its 14th Five-Year Plan (2021-2025). In January 2021 China announced its construction of the world's first integrated quantum communication network spanning over 2,800 miles, longer than the distance from New York to Los Angeles; this network has been further extended and more than doubled to 10,000 km (over 6,600 miles) by Oct 2022.)

* America retains its position of dominance in the semiconductor industry, which it has held for almost half a century. But China's decades-long campaign to become a semiconductor powerhouse has made it a serious competitor that may catch

up in two key arenas: semiconductor fabrication and chip design.

* The US has seven of the ten most-valuable life sciences companies, but China is competing fiercely across the full biotech R&D spectrum. Chinese researchers have narrowed America's lead in the CRISPR gene editing technique and surpassed it in CAR-T-cell therapy.

* Though America has been the primary inventor of new green energy technologies over the past two decades, today China is the world's leading manufacturer, user, and exporter of those technologies, cementing a monopoly over the green energy supply chain of the future. Consequently, America's green push relies on deepening its dependence on China…"

Prof Graham Allison et al conclude: "In sum, although the U.S. has led the past half century of technological innovation and still retains dominance in several other technological fields, China has emerged as a serious peer competitor in the foundational technologies of the 21st century whose applications promise to be transformative in arenas from intelligence and military power to economic growth and governance."

According to the September 2022 report of the Special Competitive Studies Project (SCSP), the US must win against China on three technology battlegrounds: microelectronics, 5G, and AI.

"In our judgment, China leads the United States in 5G, commercial drones, offensive hypersonic weapons, and lithium battery production," the SCSP has reported.

"The United States has modest leads in biotech, quantum computing, commercial space technologies, and cloud computing, but these could flip to the Chinese column. In the AI competition, the United States has a small lead with China catching up quickly across the AI stack…"

According to the SCSP report, the 2025-2030 period represents a "critical window where tech needs and strategic competition will come to a head in the contest…"

P23 4763b words 30.10.2022 16:41
11.11.2022 12:07

The US Department of Commerce has positioned itself at the forefront to thwart China's challenge to America's competitiveness.

"We believe there are three families of technologies that will be of particular importance in the coming decade: first, computing-related technologies, including microelectronics, quantum information systems, and artificial intelligence (AI); second, biotechnologies and biomanufacturing; and third, clean energy technologies," Commerce Secretary Gina Raimondo said November 2022.

"We will continue to take action to protect our advantage and maintain as large a lead as possible in these foundational technologies…" 06.02.2023 16:18

On the $39 billion Chip Fund, she said on 23 Feb 2023 the return on investment (ROI) is in protecting US national security by returning to manufacturing the world's most advanced chips in America and safeguarding its economic future.

According to the report by the Massachusetts Institute of Technology (MIT) released on 28 February 2023, the US needs more than the funding in the CHIPS and Science Act of August 2022 to stay ahead of China. (Daniel Howley, Technology Editor, **yahoo! news** 28 February 2023)

"While the CHIPS Act is certainly helpful, it doesn't address the litany of other shortfalls that are hurting America's competitiveness," Neil Thompson, director of the MIT FutureTech Research Project, told Yahoo Finance.

According to the MIT report, a 2019 survey of 120 US organizations including universities, national labs and federal agencies, found that 79% of respondents believe China is improving its <u>ability to compete</u> at a better rate than America.

According to Thompson, the US needs to do more than simply build chips and restrict China's access to them. The US must buck up, develop new kinds of computer algorithms that help businesses improve their overall efficiency, as well as stay at the forefront of <u>advanced computing</u>.

If the government and businesses can come together in the US, the country should be able to hold on to its lead in the tech industry, Thompson said. If not, China will likely grab the lead and its businesses and military will supplant the US as the top dog on the international stage.

"America's lead in <u>advanced computing</u> is nearly gone, putting in jeopardy the many dividends that leadership has historically provided," Neil Thompson has co-written with Chad Evans, Executive Vice President of the Council on Competitiveness, and Daniel Armbrust, co-founder and director of Silicon Analyst, founded in 2015 to incubate semiconductor startups. (Georgetown Public Policy Review 2023 February 28)

"<u>Advanced computing</u> is fundamental to national security, from intelligence gathering and military power, to the cybersecurity of financial markets and critical infrastructure...

"Concurrently, our survey (of over 250 biggest computing users in the US, November-December 2019) shows that China has closed the computing capabilities gap with America... However, the <u>near-parity</u> hides a more alarming outlook for the future: 79% of American respondents believe their Chinese competitors are improving capabilities faster..."

According to the March 2023 report by the Australian Strategic Policy Institute (ASPI), Canberra-based defence and strategic policy think tank and "critical tech tracker", China is beating the US in 37 of 44 technologies which are likely to propel innovation, growth and military power in coming decades, including artificial intelligence (AI), robotics, biotechnology, advanced manufacturing, and quantum technologies. (**ABC**, Stephen Dziedzie 02 March 2023)

The ASPI report says the US is the leading innovator in only seven technologies -- including quantum computing and vaccines -- and ranks second to China in most other categories.

ASPI reports that China is particularly dominant in research for the defence, security and space sectors, often producing more than five times as much high-impact research as its closest competitor -- which is almost always the US.

According to Daniel Hurst's report in **The Guardian** Thu 2 March 2023, the ASPI findings were based on "high impact" research in critical and emerging technology fields, focusing on papers that were published in top-tier journals and were highly cited in subsequent research.

To quote **HELSINKI TIMES** 23 April 2023: "**China is dominating** the global race for future power, with the country establishing a significant lead in high-impact research across the majority of critical and emerging technology domains, according to a report from the Australian Strategic Policy Institute (ASPI). The report, called the Critical Technology Tracker, examines 44 critical technologies spanning defence, space, robotics, energy, biotechnology, artificial intelligence (AI), advanced materials and key quantum technology areas..."

China leads many technologies in defence, space, robotics and transportation: in advanced aircraft engines including hypersonics (with 48.49% followed by US 11.69%), drones, swarming and collaborative robots (36.07% vis=a-vis US 10.30%), autonomous systems operation technology (26.20% vis-a-vis US 21.01%), and advanced robotics (27.89% vis-a-vis US 24.64%). US leads in small satellites (24.49%

vis-a-vis China's 17.32%) and in space launch systems (19.67% vis-a-vis China's 18.24%).

China leads all 12 technologies tracked in advanced materials and manufacturing (including nanoscale and addictive/3D); 7 of 10 technologies tracked in AI, computing and communications (advanced radiofrequency incl. 5G & 6G, advanced optical communications, AI algorithms & hardware accelerators, distributed ledgers, advanced data analytics, machine learning (ML) including neural networks & deep learning, and protective cybersecurity technologies. USA leads 3 technologies in high performance computing (HPC), advanced IC design and fabrication, and natural language processing (incl. speech and test recognition and analysis).

China leads all tracked 8 technologies in energy and environment: hydrogen and ammonia for power, supercapacitors, electric batteries, photovoltaics, nuclear waste management and recycling, directed energy (DE) technologies, biofuels, and nuclear energy.

While the US leads quantum computing, China leads technologies in post-quantum cryptography, quantum communications including quantum key distribution (QKD), and quantum sensors. China also leads in photonic sensors for sensing, timing and navigation,

In biotechnology, gene technology and vaccines, China leads synthetic biology and biological manufacturing while the US leads vaccines and medical countermeasures.

"In the long term, China's leading research position means that it has set itself up to excel not just in current technological development in almost all sectors but in future technologies that don't yet exist," **HELSINKI TIMES** has commented.

"In the more immediate term, China's lead, coupled with successful strategies for translating research breakthroughs to commercial systems and products that are fed into an efficient manufacturing base, could allow China to gain a stranglehold on the global supply of certain critical technologies…"

ASPI has reported" "Our research reveals that China has built the foundations to position itself as the world's leading science and technology (S&T) superpower, by establishing a sometimes stunning lead in high-impact research across the majority of critical and emerging technology domains…"

P37/108 23688 words 25.04.2023 20:25
P102 22215 words 03.03.2023 04:10
25.04.2023 00:39 12;05.2023 03:43

4

Innovation at the core of the US-China techno competition

President Xi Jinping has declared (May 2021), "Technological innovation has become the main battleground of the global playing field, and competition for tech dominance will grow unprecedentedly fierce..."

China's 14th Five-Year Plan (2021-2025) emphasizes the need to "develop indigenous capabilities, decrease dependence on foreign technology, and advance emerging technologies." The 14FYP also identifies key performance indicators, sets deadlines for outcomes, and holds provincial and local governments accountable for delivering results and meeting set targets.

"The link between technology, innovation, national security, and international power is now widely recognized," James Andrew Lewis, senior vice president at the Center for Strategic & International Studies (CSIS), Washington, D.C., has reported on "Technological Competition and China", November 30, 2018.

"A country's ability to innovate and produce advanced technologies provides economic strength, military power, and an intangible benefit of perceived leadership..."

An expert on international economic law and comparative international law, Professor Anthea Roberts of Australian National University has written on "The U.S.-China Trade War Is A Competition for Technological Leadership" (LAWFARE May 21 2019). Her article has been co-authored with two PhD candidates: Henrique Choer Moraes at Leuven Center for Global Governance Studies (Belgium), and Victor Ferguson at ANU's School of Political and International Relations.

"…Although the U.S.-China trade war (launched by Trump in early 2018) receives a lot of attention, it masks a more significant "tech war" over <u>innovation</u> in the 21st century.

"The United States is currently the world leader in technological innovation, which it has used to fuel both its economic advantage and its military predominance. It is now facing a technological challenge in the form of China, which is leading the U.S. to undertake shielding (protecting domestic technological knowledge), stifling (to inhibit strategic competitor's capacity), and spurring (to stimulate technological innovation) measures to protect its innovative edge," Roberts has written.

"Until recently, the United States was fairly dismissive when it came to Chinese innovation capacity, viewing China as a "copycat nation" that could only steal or "rip off" technological innovations. Yet China has made significant investments in research and development (R&D) in recent years, and Chinese companies have made impressive strides forward across a range of areas, including ICT (information and communications technology) and artificial intelligence (AI).

"As China seeks to move itself forward, the United States now faces an imperative to maintain its "technological supremacy." It accordingly has an interest in defending its existing technological dominance, hobbling the technological ambitions of its upcoming rival China and doubling down on its own technological advancement to ensure it retains its edge going forward…"

P25 5199 words 31.10.2022 11:52

Peter Cowhey, Chair, Dean of UC San Diego's School of Global Policy and Strategy, has proposed MEETING THE CHINA CHALLENGE: A New American Strategy for Technology Competition, November16, 2020. Cowhey has written:

"Innovation in science and technology (S&T) is a core American strength. The United States has been the undisputed global technology leader since the end of World War II (late 1945), but today, our preeminence faces three major interlinked challenges: The United States has allowed the foundations for its technological leadership to erode. It faces formidable competition from the People's Republic of China (PRC) -- a country that has deployed full state power, and sometimes used illegal means, to build an innovation system to gain on the United States. And it (the US) has overreacted to the competition challenge from China, and in doing so, is poised to damage its own innovation ecosystem, which flourishes in an environment of openness.

"To confront these challenges, the United States need a clear-eyed strategy for S & T innovation that enhances our national competitiveness and protects our national security. We must do two things now: make needed investments in and policy adjustments for our S & T base at home, and craft a new approach to global cooperation that minimizes the security risks China poses without unduly sacrificing the benefits of openness..."

In CHINA FOCUS A New Era of Chinese Technology and Innovation (a project of 21st CENTURY CHINA CENTER UC San Diego School of Global Policy and Strategy), The Editors featured on December 14, 2020 an interview with Dr Tai Ming Cheung, a professor at the School of Global Policy & Strategy, director of UC Institute on Global Conflict and Cooperation (IGCC), and a leading expert on Chinese and East Asian defense and national security affairs.

"... China's focus on indigenous innovation, a step up from a late catch-up technology power to an original innovation power, is front and center of China's global technology ambition, both in terms of existing technology and in emerging areas such as 5G, quantum, AI, nuclear fusion, etc. Xi Jinping regularly states that China is going to step up, and that the efforts to shut China off from the global innovation network are only spurring China on," Dr Cheung said.

"The shifting landscape is a good opportunity for China to wean itself off from what I call an "absorption-based model of technology development." Some concrete areas to analyze for changes are China's research and development (R&D) flow, and the training of next-generation talent in science and technology (S&T). We see that China is doubling down in these areas.

"The problem for China though is that such a technological leap takes time. When you try to develop new generations of technology like AI or Quantum, it does not happen overnight. It takes 10-15 years and it requires a whole ecosystem.

"You need human capital, financial capabilities, and enterprises. China under Xi Jinping has been building this ecosystem, what I call a national innovation system, but it will take a long time.

"The Chinese leadership has set the goals that by 2035, they will move into the top tier of S & T countries in the world, and by 2050, they will begin to challenge the U.S. for the "most advanced country" title. It looks like they are trying to reach that goal and are putting in the resources and political capital to make it happen."

In 2050 China (Springer 2021), Angang Hu et all of Tsinghua University in Beijing have written that China will "seize all opportunities to catch up, keep pace, and finally lead the world in the science and technology (S&T) revolution. It will become the world-leading science and technology (S&T) power by 2050…" 25.02.2023 05:02

In THE GRAND RACE FOR TECHNO-SECURITY LEADERSHIP (**WAR ON THE ROCKS** Commentary August 31,2022), Dr Cheung Tai Ming and Dr Thomas G, Mahnken have written:

"China's governing elite has been focused upon competing with the United States for several decades. Deepening concerns about the external security environment since the late 1990s, and especially the threat posed by the United States, have motivated Chinese efforts to <u>innovate</u>. Beijing has used the perception of threats posed by Washington as a catalyst to both deploy weapons and ramp up techno-security capabilities more broadly. Such perceptions have only grown more dire and expansive under Xi Jinping and serve to both motivate and focus the development of Chinese industrial and technological innovation in defense.

"By contrast, the United States has only recently begun to focus on the challenge posed by China. As China ramped up its efforts at innovation and military modernization from the beginning of the 2000s, U.S. assessments of these efforts were that <u>they posed little strategic threat as Chinese capabilities were far behind U.S. levels</u>. Indeed, while Beijing was focused on Washington, the United States was preoccupied by the Global War on Terror and threats emanating from the Middle East after the Sept. 11, 2001, terrorist attacks..." 05 November 2022 19:1

Research & Development (R&D)

<u>McKinsey Global Institute (MGI)</u>, July 1, 2019: China's scale in R&D expenditure has soared -- spending on domestic R&D rose from about $9 billion in 2000 to $293 billion in 2018 -- the second-highest in the world (after the US) -- thereby narrowing the gap with the United States.

However, China still depends on imports of some core technologies such as semiconductors and optical devices, and intellectual property (IP) from abroad.

China's technology import contracts are highly concentrated geographically, with more than half of purchases of foreign R&D coming from only three countries -- the US (31%), Japan (21%), and Germany (10%).

<u>Peter Cowhey</u>, UC San Diego November 16, 2020:

In 2019, China surpassed Japan to become the second largest funder of R&D in the world. In 2018, China spent $554.3 billion (PPP) on R&D, only slightly below the level (about $580 billion) spent by the United States. That same year, China's share of global R&D (26.3%) approached the U.S. share (27.6%). (Congressional Research Service 2020)

Over the next six years, Beijing plans to invest an additional $1.4 trillion of state and private funds in next-generation technologies. (**Bloomberg News** 2020)

<u>Graham Allison et al</u> Dec, 07, 2021: At the beginning of the century, America was number one in R&D expenditure, spending $270 billion in current purchasing power parity (PPP), followed by the E.U. at $180 billion. That same year China's expenditure was only 12% of America's at $23 billion. But by 2020, China rose to number two with 90% of America's expenditure (estimated $708 billion according to Congressional Research Service September 13, 2022).

<u>R&D Gross Expenditure (in billions U.S. dollars 2020)</u>

	US	China
2000	$270 bn	$30 bn
2010	$500 bn	$250 bn
2020	$640 bn	$580 bn

On its current trajectory, China will overtake the U.S. within the next decade. Indeed, although the U.S. maintains a strong position in long-term drivers of scientific development (where the U.S. accounts for 60% of global spending on <u>basic research</u> to China's 20%), China has focused intensely on turning scientific developments into economic products and now spends almost $70 billion more annually than the U.S. in experimental development. While six American companies top the list of the world's 10 most valuable tech companies, six Chinese companies lead the list of the 10 most valuable unicorns pioneering new technologies.

P29 6133 words 31.10.2022 17:53

"A nation that wishes to be a technological hegemon must also develop strategic plans and invest significant funds in research and development (R&D). In both categories (science & engineering talent and global patent race), the trends are concerning for the United States. While in 1960 the U.S. accounted for 69% of the world's R&D funding, by 2019 its share of expenditures was only about 30%. Meanwhile, China's share of global investment has risen from 5% in 2000 to 24% in 2019.

"In the case of China, these changes have correlated with national strategies to prioritize science and technology (S&T), including China's fifteen-year Medium and Long-Term Plan for Science and Technology Development (2005-2020), New Generation Artificial Intelligence Development Plan (2017), and Made in China 2025 (2015)," Joe Schuman has written in his article on "Military-Technology Competition Between the United States and China" (SFS Georgetown Journal of International Affairs April 12, 2022).

"Military-technology competition is just one component of this (US-China) competition, but the stakes could not be higher. The United States must recognize the problem and have the will to address it."

(A former civilian employee in Department of Defense (DoD)'s Joint Staff and Office of Undersecretary of Defense for Research and Engineering, Schuman is founder of Divided We Fall organization to help restore civility and bipartisanship in a bitterly and strongly divided American society.)

Patents

Graham Allison et al Dec, 07, 2021: In international patent filings, China displaced the U.S. as the top user of the Patent Cooperation Treaty (PCT) in 2019, when it filed 22% of PCT patents, compared to 0.6% in 2000. Meanwhile, the U.S. share fell from 42% to 22% during the same period. And in 2016, China overtook the U.S. as the top producer of scientific publications, now accounting for over 20% of science research output worldwide.

China's advantages begin with a central leadership that understands the stakes of the tech competition and aims for China to "enter the forefront of innovative countries" by 2035; an unprecedented national strategy for acquiring overseas technologies (through investments, talent programs, open source S&T collection, intellectual property (IP) theft, and academic espionage); competitive provincial governments that execute their strategies through local initiatives like high-tech parks; scale of funding; and data collected by companies and government in a society that prioritizes security over privacy.

China's government, laws and regulations, national strategies, and deep military-civilian fusion (MCF) are all green lights for its advance in key emerging technologies. Wherever the Chinese government can protect companies in its domestic market, support national champions through subsidies and access to government data, and enable corporations to lead, it does. As a result, China's tech ecosystem may be on par with Silicon Valley by 2025 "in terms of dynamism, innovation, and competitiveness." (Matt Sheehan, MacroPolo, October 26, 2020)

P31 6623 words 01.11.2022 11:36

On 30 August 2022 the Center for Strategic & International Studies (CSIS) held a panel discussion on "U.S. Technological Leadership and Patents: What Can the Data Tell Us?" Webinar was moderated by CSIS senior advisors Andrei Iancu and Kirti Gupta.

Intro/caveat: While patent data is widely used as a metric of national competitiveness and innovation, it is important to be mindful that patents vary greatly in quality. The volumes of patent applications can also be manipulated through government or firm incentives. And, of course, patents are only one measure of a country's innovation capacity.

The World Intellectual Property Organization (WIPO) IP Statistics Data Center compiles decades of historical data on patent applications and grants at both the international and national levels. This information, which incorporates data on foreign patent filings, shows that patent applications filed and grants awarded by China's National Intellectual Property Administration (NIPA) have been on a steady increase since 2000, with significant year-over-year increases beginning in the early 2010s (from over 400,000 applications in 2011-2012 to over 1.4 million since 2018 while US total patent applications have remained around 400,000 during 2013-2020).

Main observations:

1. Comparisons based on raw, aggregated patent data show that China has been the world's leader in patent applications and grants for nearly a decade. Where measuring numbers of patents granted for technologies of strategic significance, including semiconductors, computer technology, telecommunications, energy technology, and biotech, China appears to be the global leader or a peer of the United States.

2. What does the data purport to show?

By looking at relative global trends in patent filings, researchers and policymakers can gain some insight into where individual countries are focusing their efforts technologically and the speed with which they are transitioning toward a knowledge-based economy. Security analysts also examine changes in patent trajectories because cutting-edge technologies are often developed first for military usage. Large patent application numbers can also be loosely indicative of how attractive a country's intellectual property (IP) laws are, as inventors are incentivized to file where the protection of their ideas is best safeguarded.

WIPA data shows that Chinese filers dominate patent applications for inventions, utility models, or designs. In 2019, WIPO reported that China filed 1.4 million patents, or 43.4% of the world's total patent applications that year. This was more than twice the level of applications in the United States.

China's increase in patent applications appears to reflect its rapid industrialization. As China seeks to lead in a series of emerging technologies, many experts use patenting data to demonstrate the speed and trajectory of China's technological progress.

3. R&D intensity, which measures R&D expenditures as a percentage of per capita GDP, is another static marker that shows relative national efforts to sustain innovation. Notably, the World Bank's June 2022 data on this show the United States, Japan, South Korea, and most Western European nations greatly outperforming China in this regard. Other factors used for predicting a nation's innovation potential include educational achievement, number of working-age adults, firm valuations, and royalty payments, which recoup the cost of R&D and reveal the value that the market has put on that innovation.

Talent

Graham Allison et al Dec. 07, 2021: By total number of undergraduate university degrees in science and engineering, America was the global leader in 2000 with over 500,000 while China stood at just under 360,000. Today (2021), China graduates four times as many STEM (science, technology, engineering, and mathematics) students as the United States (1.3 million vs 300,000) and three times as many computer scientists (185,000 vs 65,000).

In international science and technology (S&T) rankings for K-12 students, China consistently outscores the US in math and science.

Of every ten computer science PhDs graduating in the US today, three are American and two Chinese. Three decades ago, only one of every twenty Chinese students studying abroad returned home. Now, four of every five do. And although America has historically benefited from its ability to attract talent from a global pool of 7.9 billion (with almost half of all U.S. Fortune 500 companies founded by immigrants or their children), the National Security Commission on Artificial Intelligence (NSCAI) recognized (2018) that "Competition for international students has accelerated… For the first time in our lifetime, the United States risks losing the competition for talent on the scientific frontiers."

Joe Schuman April 12, 2022: China has produced more STEM PhDs than the United States since the mid-2000s and is projected to triple the number of Americans receiving those degrees by 2025 when excluding international students studying in the United States.

China now boasts the top computer science and engineering university in the world, Tsinghua University, according to the **U.S. News and World Report**, along with five out of fifteen of the top computer science programs and eight out of fifteen of the top engineering programs.

China recently surpassed the United States for the top spot in the global patent race (a 200-fold increase in the past 20 years) as well as most research publications.

US NATIONAL SECURITY STRATEGY October 2022: "We also are (while boosting R&D) doubling down on our longstanding and asymmetric strategic advantage: attracting and retaining STEM talent is a priority for our national security and supply chain security, so we will aggressively implement recent visa actions and work with Congress to do more…"

Appendix: China's High Technology Sector (HTS)

"China's technology value chains are highly integrated globally," McKinsey Global Institute (MGI) reported on "China and the world: Inside the dynamics of a changing relationship", July 1, 2019.

"China has made huge strides in innovation in recent years; China is a a global force in the world's digital economy and artificial intelligence (AI) technologies. In many types of technology, it is already the largest consumer (for example, China accounted for 40% of global mobile phone sales in 2017, 64% of sales of battery electric vehicles sales, and 46% of semiconductor consumption). Access to the Chinese market has provided many high-tech players with significant growth opportunities. According to an MSC1 index, the US information technology (IT) sector makes 14% of its revenue in China.

"Because technology value chains are some of the most complex, they require the most collaboration, and, indeed, China is highly integrated in these value chains with a large share of technology global exports and imports. Consider, for instance, that in the case of integrated circuits (ICs) and optical devices, Chinese imports outstrip domestic production by a factor of five...

"We also found that Chinese supplies are able to provide 60-80% of the technologies studied (81 technologies in 11 areas), which means that China still uses inputs from multinational corporations (MNCs) in at least 20-40% of cases..."

US Senator Marco Rubio has written: "High-end goods made by advanced manufacturing were the very products that America was supposed to make more of due to our competitive advantages in talent and capital. Instead, these products are increasingly being captured by China..." (INTRODUCTION in MADE IN CHINA 2025 AND THE FUTURE OF AMERICAN INDUSTRY 2019)

"China has made gains vs. The U.S. in high-value sectors by various measures. Importantly, this has occurred in global export markets, which involve large scale and competition. China has significantly increased its export share of global markets since 2001 (joining the World Trade Organization) and aims for continued growth by this measure with the "Made in China 2025" plan," the American report has noted.

"Though its goals and progress vary, China has demonstrated clear success in information technology (IT), shipping, and energy and power generation, while investing in large-scale projects in aerospace, vehicles, and robotics.

"China's arrival at the technological frontier in some industries has made its next stage development something other than merely "catching up" to developed economies like the U.S. MIC2025 can be understood as both a result and cause of this move. By one translation, the plan makes this priority clear in its own words:

"Manufacturing is the main pillar of the national economy, the foundation of the country, tool of transformation and basis of prosperity. Since the beginning of industrial civilization in the middle of the 18th century (aka the First Industrial Revolution), it has been proven repeatedly by the rise and fall of world powers that without strong manufacturing, there is no national prosperity. Building internationally competitive manufacturing is the only way China can enhance its strength, protect state security and become a world power...

"China's manufacturing sector has maintained its rapid development and has built an industrial system that is both comprehensive and independent. It has greatly supported China's industrialization and modernization and significantly enhanced the country's overall strength (aka comprehensive national power). It has supported China's position as a world power.

"However, compared with the advanced economies, China's manufacturing sector is large but not strong, with obvious gaps in innovation capacity, efficiency of resource utilization, quality of industrial infrastructure and degree of digitalization. The task of upgrading and accelerating technological development is urgent... We will strive to transform China into the global manufacturing leader..." (State Council of the People's Republic of China, Made in China 2025: Charting the 10-Year Transformation of Chinese Industry, trans. IoT, Beijing, China, 2015)

"... China's high-technology sector (HTS) is increasingly integrated in the global economy. With the exception of technologies that have a military and intelligence application, which are generally off-limits to foreign investment and partnership, the globalization of technological innovation is set to continue.

"A wide range of actors, including private and public companies, governments, universities and research institutes have played a significant role in the innovation of these technologies," Dr Jue Wang, Associate Fellow, Asia-Pacific Programme (based in Holland), has written on "The Implementation of Global Technological Innovation in US-China Strategic Competition", CHATHAM HOUSE Nov 2019.

"For late entrants in the HTS, such as China, the global technology market is vital as it enabled the Chinese government and companies to buy advanced technologies that cannot be developed and produced at home. It also helps domestic developers advance their skills and techniques in technological innovation to meet international standards.

"Chinese actors have taken pro-active measures to engage in the globalization of technological innovation, including:

1. Exports and imports of technological goods and services;
2. Cross-border investments in technology companies and research and development (R&D) activities;

3. Cross-border R&D collaboration; and
4. International techno-scientific research collaboration.

"Each of these actions has generated technological gains for Chinese actors. China is now home to several world-leading tech conglomerates and is <u>the largest global high-tech products exporter</u>...

"China's National Bureau of Statistics (NBS) data show that China's has had a trade surplus of high-tech products since 2000. And, according to the World Bank, since 2004 it has been the largest exporter of high-tech products globally.

"As of 2017, the total value of China's high-tech exports was $654 billion, more than triple that of the second largest exporter, Germany. This clear increase in China's economic competitiveness has pressured traditional high-tech leaders like Germany, the US and Japan. However, the vast majority of China's high-tech exports are only assembled in China and most of the profits in these industries go to companies in the US, Europe, and developed Asian economies. Moreover, China still relies on developed economies for the most advanced high-tech products with higher values..."

8488 words 01.11.2022 18:11 11.11 11.11.2022 13:31
06.02.2023 17:17
P53/106 23.04.2023 03:15 20:45
P54/116 12.05.2023 04:55

5

The US Wages Techno-Economic Offensive Against China

DEPARTMENT OF COMMERCE Bureau of Industry and Security Wednesday, August 1, 2018

Federal Register/Vol. 83, No. 1481 15 CFR Part 744

This rule amends the Export Administration Regulations (EAR) by adding forty-four entities (eight entities and thirty-six subordinate institutions) to the Entity List. The entities that are being added to the Entity List have been determined by the U.S. Government to be acting contrary to the <u>national security or foreign policy interests</u> of the United States. These entities will be listed on the Entity List under the destination of China...

This rule is <u>effective August 1, 2018</u>...

The principals:

1. China Aerospace Science and Industry Corporation (CASIC) Second Academy + 13 Subordinate Institutions
2. China Electronics Technology Group Corporation 13th Research Institute (CETC 13) + 12 Subordinate Institutions
3. China Electronics Technology Group Corporation 14th Research Institute (CETC 14) + 2 Subordinate Institutions
4. China Electronics Technology Group Corporation 38th Research Institute (CETC 38) + 7 Subordinate Institutions
5. China Electronics Technology Group Corporation 55th Research Institute (CETC 55) + 2 Subordinate Institutions
6. China Tech Hi Industry Import and Export Corporation

7. China Volent Industry, and

8. Hubei Far East Communication System Engineering.

Subsequently, Huawei & affiliates (69 entities) were similarly corralled into the Entity List in May 2019 for involvement in activities "contrary to the national security or foreign policy interests of the United States".

"The United States recognizes the reality of China's tech threat and the imperative to maintain a competitive edge," Mercy Kuo commented 06 May 2021 (ISP Italian Institute for International Political Studies China Int. Strategy Rev. 2 published 24 June 2020).

"US legislation -- including the Foreign Investment Risk Review Modernization Act of 2018 (FIRMA), the Export Control Reform Act of 2018 (ECRA), and the Holding Foreign Companies Accountable Act of 2020 (HFCAA) -- reflects Washington's attempts at limiting China's access to technology and capital markets.

"FIRMA requires heightened scrutiny of foreign investment in "critical US technologies". ECRA authorizes the US Department of Commerce to implement export controls on "emerging and foundational technologies". HFCAA requires certain issuers of securities to verify they are not owned or controlled by a foreign government.

"The Biden administration recently affirmed Executive Order 13959, "Addressing the Threat from Securities Investments that Finance Communist Chinese Military Companies", de-listed and de-indexed 31 firms trading in U.S. markets that are owned or controlled by the Chinese military.

The cascading effect of the legislation has sped up China's push to indigenize supply chains and globalize China-led technology standards..."

Ashley Feng and Lorand Laskai reported on the US-China Tech Competition (**Defense One** September 3, 2019) that the U.S.

Department of Commerce had halved the number of licenses that let U.S. companies assign Chinese nationals to sensitive technology and engineering projects.

However, Assistant Secretary Dr. Christopher Ashley Ford's accusation against Chinese companies such as Huawei, ZTE, AliBaba, Tencent, and Baidu of cooperating and collaborating with China's defense and security authorities can be equally, if not more, lodged against US military contractors such as Northrop Grumman, Raytheon, Boeing, Lockheed Martin, etc.

"In June (2019), the U.S. Department of Commerce added five Chinese companies involved in the Chinese government's supercomputing program to the Entity List, which prevents U.S. companies from doing business with them. Not just another shot across the bow in the U.S.-China trade war, these additions reflect the national security ramifications of selling commercial processors to companies linked to the Chinese military," Feng and Laskai reported.

"Unlike the President's trade war, support for this new, expansive definition of national security and technology is largely bipartisan and likely here to stay. Throughout Washington, concerns about Xi Jinping's civil-military fusion (CMF) drive and renewed emphasis on acquiring dual-use technologies has opened the door for more enforcement against Chinese entities seen as vectors for Beijing's high-tech push…"

U.S. Department of Commerce Press Release April 18, 2021 Commerce Adds Seven Chinese Supercomputing Entities to Entity List for their Support to China's Military Modernization and other Destabilizing Efforts:

The Department of Commerce's Bureau of Industry and Security (BIS) has added <u>seven Chinese supercomputing entities</u> to the Entity List for conducting activities that are contrary to the national security or foreign policy interests of the United States.

Today's final rule (under Section 744.II of Export Administration Regulations) adds the following entities to the Entity List: Tianjin Phytium Information Technology, Shanghai High-Peformance Integrated Circuit Design Center, Sunway Microelectronics, the National Supercomputing Center Jinan, the National Supercomputing Center Shenzhen, the National Supercomputing Center Wuxi, and the National Superconducting Center Zhengzhou.

These entities are involved with building supercomputers used by China's military actors, its destabilizing military modernization efforts, and for weapons of mass destruction (WMD) programs.

U.S. Secretary of Commerce Gina M. Raimondo's released statement: "Supercomputing capabilities are vital for the development of many -- perhaps almost all -- modern weapons and national security systems, such as nuclear weapons and hypersonic weapons. The Department of Commerce will use the full extent of its authorities to prevent China from leveraging U.S. technologies to support these destabilizing military modernization efforts."

On 9 August 2022 President Joe Biden signed into law the CHIPS and Science Act of 2022, known colloquially as CHIPS+, with $52 billion to help secure US leadership in the global semiconductor industries, restore America's chip manufacturing capacity (now 12% of global from about 40% in 1990), reduce reliance on foreign-made semiconductors, particularly the most advanced chips used in cutting-edge civilian and military technologies, and revive high-tech innovation in domestic manufacturing and R&D.

BBC News reported 9 August 2022 that Biden signed the CHIPS and Science Act with $280 billion commitment to high-tech manufacturing and scientific research amid fears the US is losing its technological edge to China.

"By enacting the CHIPS and Science Act we are making clear we believe another great American century lies on the horizon," said top Senate Democrat Chuck Schumer, who hailed the bill as a "game changer" to ensure American leadership and prosperity in the next century.

"Biden signed CHIPS and Science Act in a move to crack down on China's semiconductor supply chain. The bill includes $52 billion for chip manufacturing and research, as well as a $24 billion investment tax credit for chip plants in US," Xie Jun reported Aug 09, 2022 **Global Times**.

"However, according to a **Bloomberg** report, to qualify for subsidies, companies must not expand their semiconductor manufacturing in China for 10 years…"

On Aug 10, 2022 **Global Times** carried a wide-eyed and in-depth view by Ding Gong, a senior editor with **People's Daily** and a senior fellow with Chongyang Institute for Financial Studies at Renmin University of China. To quote:

, "The bill contains a clear intent to exclude China from international industrial and technological cooperation, the standard-setting system, and the supply chain of high-tech products. The bill explicitly prohibits US companies that receive development support from investing in China's related infrastructure…

"Washington will do its best to squeeze China's living space in the global research development and market system, and block the path of Chinese manufacturing and capital to high-end development and the global market …

"The US is attacking China's strengths in its development, its all-encompassing, large-scale made-in-China, its high-technology represented by artificial intelligence (AI) and 5G; its expansion into the global market, and its advancement into the industrial standards system.

"For China, it is imperative that we follow our strategic arrangement, handle our priorities and break through with our strengths…

"The US has the high-end equipment and R&D capabilities to engage in chip alliances. We have the market and a large-scale, all-round manufacturing system, and can also engage in cooperation in key industries. The more the US tries to squeeze China out of the industrial chain, the more we need to improve our ability to bond with the global industrial chain.

"At the same time, China also needs to use the power of the country to promote Chinese technology, Chinese standards, and made-in-China to the world."

In "Washington's chips assault will only hasten China's indigenous innovation", Wen Sheng, a **Global Times** editor, has written Sep 04, 2022: "The Biden administration has moved to further restrict China's access to advanced semiconductor products and technology, placing seven more Chinese entities of aerospace on its export control list late August (2022), and ordering Nvidia and AMD last week to stop providing high-end artificial intelligence (AI) and supercomputing microchips to Chinese companies in Washington's complete disregard for international trade rules…

"So far, the US department of commerce has put as many as 600 Chinese entities on its export restriction list, more than 110 of which have been added since Biden came to office (January 2021). These 600 entities are the key business players in almost all China's competitive sectors, such as telecom, information, semiconductors, aerospace, supercomputing and AI (artificial intelligence).

"Will the ruthless US assault decimate Chinese scientists' indigenous innovation? The answer is a clear no. Washington's attack will only slow down China's progress in some technology fields in a short time. But China will weather the attack and come out of it much stronger, which has been verified by China's advancements in 4G and 5G information technology, high-speed railway technology, spacecraft launching and in-orbit maneuvering and control technology…"

P46 9937 words 03.11.2022 18:36 10021 words 04.11.2022 11:31
P60/106 23288 words 23.04.2023 21:11

According to **BBC News** 7 September 2022, US tech companies that receive federal funding will be barred from building "advanced technologies" facilities in China for 10 years.

As reported by **Al Jazeera** 5 October 2022, the world's biggest drone manufacturer DJI has been blacklisted with 12 other companies for alleged ties to Chinese military.

"The Department (of Defense) is determined to highlight and counter the People's Republic of China's Military-Civil Fusion (MCF) strategy, which supports the modernization goals of the People's Liberation Army (PLA) by ensuring its access to advanced technologies and expertise are acquired and developed by PRC companies, universities, and research programs that appear to be civilian entities," DoD declared in its press statement.

More than 60 Chinese companies have been blacklisted, including semiconductor manufacturer SMIC and IT giant Huawei Technology.

"… DJI is not a military company in China, the United States, or anywhere else," DJI spokesman Adam Lisburg told **Al Jazeera**. "DJI has never designed or manufactured military-grade equipment, and has never marketed or sold its products for military use in any country. Instead, we have always developed products to benefit society and save lives. We stand ready to formally challenge our inclusion on the list…"

According to **BBC News** 7 October 2022, the US is introducing further measures to restrict sales of computer chip technology to China in a bid to hobble the country's military advances. Under new rules, the

US said it would bar US firms from selling certain chips used for supercomputing and artificial intelligence (AI) to Chinese companies.

The restrictions also target sales from foreign firms that use US equipment.

The US is engaged in an arms race with China over control of the supply of semiconductors.

The sweeping new measures will make it harder for China to obtain advanced chips for cutting-edge technologies.

Alan Estevez, undersecretary at the US Commerce Department, announced the rules, saying his intention was to ensure the US was doing everything it could to prevent "sensitive technologies with military applications" from being acquired by China.

The US has previously barred sales of technology to specific Chinese companies, such as Huawei, on national security grounds. But these measures go much further, with many of the measures aimed at preventing foreign firms from selling advanced semiconductors to China, or providing China with the tools to make advanced chips.

The measures come as the US pours billions of dollars into its domestic chip industry, moves aimed at boosting US competitiveness.

US hits China with sweeping tech export controls to limit Chinese companies' access to advanced computer chips and slow their AI progress, Demetri Sevastopulo in Washington and Kathrin Hille in Taipei reported October 7, 2022 **FINANCIAL TIMES**.

7 October 2022 US Commerce Department introduces sweeping export controls to complicate efforts by Chinese companies to develop cutting-edge technologies with military applications as well as to make it extremely hard for them to obtain or manufacture advanced computer chips and to slow their progress in AI. Measures will make it much

tougher for China to develop supercomputers with a range of military applications from modelling nuclear weapons to making hypersonic weapons.

According to Paul Triola, a China and tech expert at consultancy Albright Stonebridge, US action has marked a "major watershed" in US-China relations and in their increasingly intense technology competition.

The restrictions will also prohibit "US persons" -- American citizens and companies from providing direct or indirect support to Chinese companies involved in advanced chip manufacturing. Kevin Wolf, an expert on export controls at Akin Grump, has described them as the "most significant and expansive".

"The administration's strategy is to deny China the capability to indigenize its semiconductor industry," said Martiju Rasser, a security and technology expert at think-tank Center for A New American Security (CNAS). "If the US is successful, this causes a huge problem for Beijing's strategy to be a world-class player…"

One chip industry executive said: "They are not just targeting military applications, they are trying to block the development of China's technology power by any means."

Ian King, Eric Martin, Jenny Leonard, Debby Wu & Ryan Vlastelica reported from San Francisco Oct 7, 2022 (**THE EDGE MARKETS/ Bloomberg** October 08, 2022): The Biden administration announced new restrictions on China's access to US semiconductors technology, escalating tensions between the two countries and adding fresh complications to an industry reeling from a slump in demand.

The measures are aimed at stopping Beijing's push to develop its own chip industry and advance the country's military capabilities. They include restrictions on the export of certain types of chips used in artificial intelligence (AI) and supercomputing and also tighten rules on the sale of semiconductor manufacturing equipment to any Chinese company.

Washington is looking to ensure that Chinese companies don't act as a conduit for the transfer of technology to their country's military -- and that chipmakers there don't develop the capability to make advanced semiconductors themselves.

US Commerce Department has also put a raft of restrictions on supplying US machinery that's capable of making advanced semiconductors. It's going after the types of memory chips and logic components that are at the heart of state-of-the-art designs.

The US is home to the biggest block of companies that design vital electronic components and provide the complex machinery needed to manufacture them -- but other regions have capabilities that could undermine some of the government's efforts.

Chipmaking gear restrictions cover production of

* Logic chips using so-called nonplanar transistors made with 16nm technology or anything more advanced
* 18nm dynamic random-access memory (DRAM) chips
* Nand-style flash memory chips with 128 layers or more

"Visions of a technologically ascendant China keep American strategists up at night. They see the contours of a surveillance state implementing the will of President Xi Jinping by algorithmic edict at home and projecting computing power abroad. To erase those contours for good, on October 7th President Joe Biden's administration announced the most sweeping set of export controls in decades," **The Economist** reported in its distinctive prose on Oct 13 2022.

"The new rules cut off people and firms in China from many advanced technologies of American origin and from products made using these.

The list includes chips used for artificial intelligence (AI), software to design advanced chips and the machine tools to manufacture them. Selling such things to China is now barred without explicit permission from America's government. Rulebreakers risk being cut off from American tech themselves…"

Jeff Pao reported in **ASIA TIMES** November 3, 2022 that on October 7, 2022, the Bureau of Industry and Security (BIS), Department of Commerce, added Yangtze Memory Technologies Co (YMTC), China's top memory chipmaker, and 30 other Chinese "entities" to an Unverified List (UVL) -- to quote BIS "on the basis that BIS was unable to verify their *bona fides* because an end-use check could not be completed satisfactorily for reasons outside the U.S. Government's control" (15 CFR Part 744 RIN 0694-A151).

BIS said the named companies would face restrictions to purchase US products. It has also banned US citizens from supporting the development or production of integrated circuits (ICs) at certain chip fabs (fabrication plants) in China, without licenses, from October 12, 2022.

Yangtze Memory Technologies Co (YTMC) is widely regarded as being the best bet China has of breaking through into the front ranks of the semiconductor industry; it has made marked progress with advanced products for chip-based storage.

Mark Magnier in New York reported 12 Oct, 2022 **South China Morning Post** that US and China both face tough struggles as war for chip supremacy intensifies following American export restrictions.

* Washington imposes sweeping new rules targeting China but faces major challenges in bid to regain its global primacy in semiconductor manufacturing
* Asian rivals far outpace the US in making the advanced chips crucial for fighter jets, drones and cutting-edge military hardware

Restrictions announced last week (7 October 2022) on Chinese access to US semiconductor technologies raise the stakes significantly in the US-China tech war, but Washington's bid to regain <u>chip manufacturing primacy</u> and slow China's military and economic rise faces huge challenges, industry analysts said.

THE ODDS are long at best that the US chip industry can again dominate in a field it pioneered, catch market leader Taiwan Semiconductor Manufacturing Company (TSMC) -- which makes some 90 per cent of the world's most advanced chips -- or match Samsung Semiconductor Global any time soon.

Nikkei reported November 2, 2022 (**FMT**) that the US holds 12% of the global semiconductor market, Taiwan and South Korea 20% each, and Japan 15%.

Washington anticipates that bringing allies on board with the sweeping export controls of Oct 7 will make that much more challenging for Beijing to buy or make advanced semiconductors for its rapidly expanding military.

In addition to exports of chips and chipmaking technology, Washington's curbs restrict US nationals from working at or doing business with Chinese semiconductor companies.

China's market for chip production equipment is estimated US$22 billion in 2022 -- 22% of the global, behind only Taiwan and South Korea. 05 November 2022 18:39 19:47

"Semiconductors are <u>ground zero</u> in this technological competition (with China) and central to our new investment strategy (initiated with a declared public investment of $1 trillion, including $52 billion in domestic semiconductor manufacturing)," US Commerce Secretary Gina Raimondo said November 2022. "The CHIPS and Science Act

(August 2022) marks the beginning of a new chapter in U.S. innovation when we reverse that decline (in US chip capacity) and ensure that the United States retains its leadership in the technologies and industries of the 21st century..."

1.1.2023   12 07.02.2023 18:16 23.04.2023 19:45 12.05.2023 18:30

6

China seeks tech self-reliance & self-sufficiency

"… MIC25 (Made in China 2025) is a blueprint for high-tech "self-sufficiency" and provides the foundation for a move up the industrial value chain to achieve competitiveness in global markets," stated the 2019 report of US Senator Marco Rubio MADE IN CHINA 2025 AND THE FUTURE OF AMERICAN INDUSTRY.

"… China is implementing Made in China 2025 to promote the deep integration of informatization and industrialization and to realize the leap-forward development of its manufacturing industry. Further, China expects to maximize self-sufficiency. One of the key objectives of Made in China 2025 is to raise the indigenous proportion of the core components and materials in high-tech manufacturing to 70 per cent by 2025," Xiangming Wu of University of Macao has written in **Springer** China International Strategy Review & **Nature** 24 June 2022.

"When President Xi (Jinping) says that China's goal is "being the master of its own technologies", anxieties seem to be the natural sentiment for Washington, which believes that the U.S. faces real and threatening competition from China's plans to become a leader in high-tech innovation. Made in China, in addition to the Belt and Road Initiative (BRI) and the Asian Infrastructure Investment Bank (AIIB), is believed to substantially threaten the global hegemony of the US…"

According to Xiang Ligang, Director General of Beijing-based Information Consumption Alliance (ICA), China aims to raise its chip self-sufficiency rate from around 30% in 2019 to 70% by 2025. (**GLOBAL TIMES** Apr 29, 2023)

By 2035, China will be fully self-sufficient in semiconductors as well as technologically self-reliant.

With about 85% self-sufficiency, "the global share of the Chinese semiconductor industry would grow from 3% to more than 30%, displacing the US as global leader," Antonio Varas and Raj Naradarajan, two top executives of Boston Consulting Group (BCG) have reported March 2020.

"An extrapolation of the ratio between the global market of Chinese companies and the weight of China's domestic market observed in other technology sectors indicates a potential 35% to 55% of global share for the Chinese semiconductor industry in the long term…" 08.06.2023 04:35

Bloomberg News on 16 October 2022 highlighted in "China's Xi Vows Victory in Tech Battle After US Chip Curbs":

* Nation to speed up technology self-reliance drive, leader says
* US aims to stop China from getting capabilities seen as threat

P55 11962 words 05 November 2022 21:00 p70/114 25363 words 08.05.2023 04:37

"President Xi Jinping said at the opening of the meeting (the Chinese Communist Party's 20th National Congress 16 October 2022), held every five years, that the country must "regard science and technology (S&T) as our primary production force, talent as our primary resource, and innovation as our primary driver of growth"," Smriti Mallapathy reported in **nature** 27 October 2022.

NIKKEI Asia reported November 21, 2022: "Chinese President Xi Jinping, who recently began a historic third term as party leader, has vowed China "will win the battle in key core technologies." Beijing, he said, will accelerate the programs related to "independence and self-reliance" over the next five years in order to compete with the U.S. in the field of high technology (HT)…" P59 17264 words 25 November 2022 19:20

The furious and relentless offensive by the US government to smother, supress and subdue its fast-rising Asian challenger has further stimulated and fully resuscitated the historical Chinese instincts of fearless struggle for survival.

With the vibrant heading "Huawei's industrial 5G takeoff…", Scott Foster has reported November 2, 2022 **ASIA TIMES** that one of America's early prime targets has literally re-invented itself to reassert its technological innovative spirit as well as commercial entrepreneurship. Foster's comprehensive report throws light on milestones of the very recent past and the near future:

Huawei's sales of 448.8 billion yuan (US$61 billion) and profits of unaudited 6.1% on its "main business" in the first nine months of 2022 "are showing signs of recovery after more than three years of U.S. sanctions" while "its industrial 5G is taking it far behind smartphones as a shaper of the global economy".

Huawaei's new industries include private 5G networks for industrial facilities, electric vehicles (EVs), autonomous driving, and cloud computing.

"In short, Huawei is becoming considerably more sophisticated than the 5G smartphone and network equipment maker that incurred the wrath of Donald Trump (May 2019)," Foster commented.

At the company's 13th annual Global Mobile Broadband Forum held in Bangkok 25-26 October 2022 and co-hosted by mobile industry association GSMA and GTI, Rotating Chairman Ken Hu pointed out in Day 1 keynote address that while consumer services still generate the largest share of telecom revenue, "B2B 5G applications are also becoming a new engine for carrier revenue growth, producing considerable value in industries like oil and gas, manufacturing, and transportation.

"These applications are not only <u>innovative</u> -- they're generating real commercial value for carriers. In 2021, for example, Chinese carriers brought in over CNY3.4 billion (roughly USD500 million) in new revenue from more than 3,000 industrial 5G projects. What's more, these projects also generated 10 times that amount from related data and integrated ICT (information and communications technology) services...

"With large bandwidth and low latency, 5G can be <u>integrated with cloud (computing) and AI to provide entirely new services</u> for consumers and businesses alike... presenting an opportunity for carriers to go beyond connectivity and move into cloud services and system integration."

On Day 2 of the Forum, Executive Director David Wang spoke about <u>5.5G</u>, which he called "the foundation of the future" and "the next milestone... on the path to <u>the intelligent world</u>". Wang said, "5.5G will deliver 10Gbit/s experience, support hundreds of billions of connections, and help us achieve <u>native intelligence</u>..."

According to Wang, 5.5G has made great progress over the past two years, "and three things have become clear:

"First, the standardization of 5.5G has been initiated and is right on track, making it more than just a <u>vision</u>.

"Second, the industry has made <u>breakthroughs in key technologies for 5.5G</u>, and ultra large bandwidth and ELAA (Enhanced Licensed Assistance Access) can deliver 10Gbit/s experience.

"Third, the industry has <u>a clear vision for the IoT (Internet of Things) landscape</u>..."

Wang concluded: "As our standards, spectrum, products, and ecosystem mature, 5.5G will become a reality, allowing even more applications to emerge. Multi-sensory interactions will transform the way we communicate. Intelligently connected vehicles are set to become a third mobile space and see wide adoption, while intelligent connections across industries will lead to the dissolution of information silos, driving industrial upgrade.

"A new generation of innovative applications is now emerging, and our vision for the intelligent world is becoming clearer. That's why all industry players need to work together towards the exploration and creation of these applications."

During the Forum, Huawei, China Mobile and Chinese appliance maker Midea launched what they call the first fully-connected 5G smart factory in 3C (Computers, Communications, and Consumer Electronics) industry.

Equipped with Kuka robots, the plant can assemble a washing machine in 15 seconds. Double shipment, halve inventory, and reduce labor costs by 30%.

China has installed about 60% of the world's 5G base stations and has more than one billion 5G users.

According to Ken Hu, more than 230 carriers worldwide have launched 5G services and over 3 million 5G base stations have been set up by October 2022.

According to ABI Research, China also leads the world in private 5G networks with over 30% of the revenue-based market.

GlobalData has rated Huawei's mobile core portfolio as the world's strongest for the fourth year running.

"Western countries -- led by the US -- have implemented all-round containment, encirclement and suppression against us, bringing unprecedentedly severe challenges to our country's development," President Xi Jinping said on 7 March 2023 at the People's Political Consultative Conference on the sidelines of the National People's Congress in Beijing. (**Xinhua** 7 March 2023)

"In the coming period of time, the risks and challenges that we face will only increase and intensify even more," Xi said.

Xi has been pushing local officials to bolster their economic resilience against international sanctions by developing self-reliance in critical areas such as technology.

(Eryk Bagshaw March 7, 2023 **The Sydney Morning Herald**)

"Senior policymakers increasingly see tech-sufficiency as a life or death issue as the US steps up efforts to stymie China's tech capacity," Chinese research firm Trivium has reported.

"That means that China will employ any and all means to master critical technologies at home…"

P71/104 22635 words 30.03.2023 04:37
P59 12659 words 06.11.2022 13:13 13:15 12834 words 06.11.2022 17:31
07.02.2023 18:36
1.1.2023:36 12.05.2023 21:12

7

Will the US and China decouple?

"Locked in a long-term competition around advanced technologies, the US is using outdated policy tools to slow China's rise as a technology power. A worst case scenario: decoupling of the two countries' technology, financial, and economic sectors" -- SYNOPIS of the article by Paul Triolo "US-China Competition: The Coming Decoupling?", contributed to S. Rajaratnam School of International Studies (RSiS) in Singapore, and published 23 October 2019. Head of global technology policy at the political risk consultancy Eurasia Group, Triolo spent 25 years in government tracking China's rise as a technology power.

"The United States and China are now clearly locked in a competition over dominance of the technologies of the future. At the same time, some elements of key government organizations on both sides are pushing for a major decoupling of very intertwined supply chains and value generation ecosystems that have developed over more than 30 years," Triolo has commented.

He has also cited 4 major factors driving US policy towards China on technology issues: (1) Industrial policies and trade; (2) Technology control; (3) Supply chain security; and (4) Struggle for dominance of the "technologies of the Future".

Triolo has written: "The ultimate prize in the view of some hardliners on both sides is power in tomorrow's international order. Control of the international marketplace is also at stake, because advanced technology will help run the global economy. Technology supply chains and dependencies will anchor future alliances and trading relationships..".

And he has further commented: "Also, <u>leadership in technology</u> will give an edge on the national security front..."

Both adversaries and their respective supporters should ponder deeply on Triolo's insightful concluding comment: "A permanent and deep US-China technology rift would substantially restrict beneficial exchanges of technology, investment, and human capital that drive new technology advances -- this will hurt <u>innovation</u> globally, leaving everyone worse off." P61 13150 words 06.11.2022 18:41

In Chatham House Research Paper November 2019 on US-China Strategic Competition, Marianne Schneider-Petsinger, a research fellow in the US and the Americas Programme at Chatham House, wrote in Chapter 2 on Behind the US-China Trade War: The Race for Global Technological Leadership.

"... President Trump's willingness (in 2019) to use Huawei's leverage in the trade talks with China has blurred the lines between US legal processes, the US-China trade war, and the quest for technological leadership. Nonetheless, the export blacklisting of Huawei shows that the US's aim is not simply about reducing the trade deficit, but about <u>decoupling</u> from China," Schneider-Petsinger has written incisively.

"<u>Decoupling</u> from China's economy would amount to the US inflicting economic self-harm. The US technology sector is deeply intertwined with China's. A fencing off would have adverse consequences for the innovation and competitiveness of US firms and raise costs for American consumers. Thus, a balance must be struck between protecting important US technology sectors for national security reasons, protecting FDI (foreign direct investment) in the US, and protecting US innovation..."

In their March 2020 paper on "How Restrictions to Trade with China Could End US Leadership in Semiconductors" submitted by two top executives of Boston Consulting Group (BCG), Antonio Varas and Raj Naradarajan have reported:

"Beyond the financial impact, our analysis also reveals a risk that shutting US semiconductor companies (with export ban and technology decoupling) could trigger dramatic structural change in the industry with deep irreversible implications for US economic competitiveness and national security. If the global share of US semiconductor companies (48%) slips to approximately 30%, the US will cede its long-standing global semiconductor leadership to either South Korea or China…"

Moreover, the substantial erosion of revenue, loss of leadership status, declining competitiveness and diminishing national security will be accompanied by the possible risk of the US becoming a net importer of chips. To quote: "More fundamentally, the US could be at risk of having to depend to a significant degree on foreign suppliers to serve its own domestic demand for semiconductors…"

Varas and Naradarajan have further observed and noted: "And with a projected 30% to 60% reduction in annual R&D investment, the US industry might no longer be able to meet the future needs of the US defense and national security systems…"

08.05.2023 20:52 21:02 p77/114 25575 words

Peter Cowhey, Chair/Dean of UC San Diego's School of Global Policy and Strategy, has delivered a similar message and cautionary note.

"While recognizing the challenges posed by the People's Republic of China, trying to shut China off from the United States and the global economy ultimately harms the United States. To remain truly competitive, U.S. firms need to operate at scale throughout the world; localize R&D to meet the needs of diverse, fast-moving markets, and hire the best talent wherever it is available," Cowhey has written November 18, 2020 on MEETING THE CHINA CHALLENGE: A New American Strategy for Technology Competition (UC San Diego's 21st Century China Center).

"While we hope that radical <u>decoupling</u> will never be necessary, and understand that such a step would have dire consequences for the global and American <u>innovation systems</u>, we would be foolish to ignore the possibility that it may become unavoidable. Unless and until such a decision is made, the role of the scientific and tech (S&T) community should be to pursue worldwide collaboration in accordance with practices that mitigate the risks from <u>openness</u>..."

"The world is such an integrated place," Helge Berger, head of the IMF's China mission, said in an interview with **Bloomberg** in New York 16 April 2021. (**TheStar** Malaysia 16 Apr 2021) "If you stop exchanging knowledge across countries or borders, you will ultimately pay a price, and this could be fairly high."

According to IMF research estimates, a technological US-China decoupling could cost about 5% of global GDP, more than ten times that (about 0.4%) of their trade war with tariffs on over $300 billion in annual imports. 11.11.2022 17:48

Semiconductor Industry Association (SIA) has reported in White Paper TAKING STOCK OF CHINA'S SEMICONDUCTOR INDUSTRY 13 July 2021: "A recent study (by US Chamber of Commerce) found expanded -- or even complete -- <u>decoupling</u> from China would lead to an 8% to 18% decrease in American chip firms' global market share, resulting in significant cuts to R&D and capital expenditures, as well as up to 124,000 lost U.S. jobs, eventually leading to the diminution of U.S. global leadership in the industry. Maintaining access to the Chinese market for U.S. semiconductor firms for the sale of general-purpose (GP) commercial chips is critical to ensuring U.S. competitiveness..."

10.05.2021 20:18

"With little fanfare or public debate, America has embarked on one of its most difficult and dangerous international challenges since the Cold War. The task: reversing decades of economic and technological integration with its chief rival, China," Jon Bateman, a senior fellow in the Technology and International Affairs Program of the Carnegie Endowment for International Peace, has written in <u>Opinion</u>, **POLITICO MAGAZINE** 12/5/2022 4:30 AM EST.

"This <u>technological decoupling</u>, if done selectively, will help from unfair competition, and push back Beijing's human rights abuses. But if decoupling goes too far, it will drag down the U.S. economy, drive away allies, stymie efforts to address global crises like climate change, and increase the odds of a catastrophic war..." 30.12.2022 23:33

Following the October 2022 announcement by the US Bureau of Industry and Security (BIS) of new extraterritorial limits on the export to China of advanced semiconductors, chip-making equipment and supercomputer components, Bateman has further commented in another article "<u>Biden Is Now All-In on Taking Out China</u>" (**Foreign Policy** Winter 2023):

"... China's technological rise will be slowed at any price...

Now the United States has gone all-in -- wagering like never before and placing its cards on the table for all to see. The decisive American gamble: to openly block China's path to become an advanced economic peer, even at significant risk to U.S. and allied interests. Bigger U.S. moves are probably coming in the future. But for now, Washington must next wait to see how others play their hands."

P66 18136 words 11.01.2023 20:59

"... The United States should seek to foster strategic dependencies by China on U.S. technology. In the U.S.-China tech competition, the leverage the United States used to kneecap Huawei is a priceless strategic advantage. Yet the Biden administration's new chip restrictions (of October 2022) undermine the United States' long-term position. Because U.S. export controls will only have a temporary effect as global markets adapt over time, the United States is better off holding this leverage in reserve for now.

"Artificial-intelligence (AI) capabilities are rapidly advancing, and computing hardware is becoming an even more valuable input for AI power," Paul Scharre, vice president and director of studies at the Center for a New American Security (CNAS), has written in **Foreign Policy (FP)**, January 13, 2023 8.00 am.

"The deep learning revolution is only a little over 10 years old, and the second decade looks to be even more dramatic than the first. The United States will be in a stronger position for the changes ahead if it retains the ability to deny China access to powerful AI capabilities, if necessary.

"Using narrower export controls today could still deny China's military access to advanced chips while keeping Chinese commercial companies reliant on foreign supplies. Keeping China dependent on U.S. technology is a stronger strategy than decoupling and will give the United States more control over China's access to advanced technology in the long run."

16.01.2023 01:15

US Commerce Secretary Gina Raimondo said in November 2022: "We want to promote trade and investment in areas that do not threaten our core economic and national security interests or human rights values. American trade between our two countries has grown exponentially from $4.7 million in 1972 (President Nixon's visit to China) to more than $750 billion today (2022)…"

Two top executives in the Boston Consulting Group have pointed out that trade restrictions and technological decoupling could terminate US leadership in the global semiconductor industry. Moreover, China could emerge as the new leader in the chip world.

And, to quote from the CHIPS for America report issued by CHIPS Research and Development Office April 25, 2023:

"It is hard to imagine a sector more important to the national and economic security of our nation than the semiconductor industry. Nearly every innovation of the future, whether for smarter transportation or better medical devices, for agricultural efficiencies or to address climate risks, or for the technology that powers our national defense capabilities, will depend on semiconductor technology…"

P86 19031 words 05.02.2023 02:55 08.02.2023 17:01 24.04.2023 20:00
P81.116 26096 words 12.05.2023 22:32

8

The nascent triad of the emerging technologies

(1) Artificial Intelligence (AI)

"As countries across the globe race to advance artificial intelligence (AI), it is crucial we get the policy environment right to enable American innovators to lead the AI revolution," Tom Quaadman has written (U.S. Chamber of Commerce August 16, 2022). Executive Vice President at the Center for Capital Markets Competitiveness (CCMC), Quaadman also serves as Senior Advisor to the President and CEO of U.S. Chamber of Commerce.

"Across the country, artificial intelligence (AI) is powering machines and computers to help us solve problems and work more efficiently. It's assisting scientists to develop vaccines and treat patients more effectively, securing our nation's networks and critical infrastructures against cyber attacks, alerting customers of bank fraud and expanding financial opportunities for underserved communities through access to credit, and much more.

"AI is rapidly changing how businesses operate -- and is <u>foundational</u> to a thriving 21st-century economy. By 2030, 70% of businesses globally expect to use AI. Around the world, AI is estimated to boost global GDP by 14% over the same period, accounting for nearly $15 trillion of economic output.

"From basic needs, such as food security and supply chain resiliency, to ensuring our nation's competitive advantage through research and development (R&D) and the intellectual property (IP) rights that underpin it, <u>AI will shape the new economic era</u>. It's no wonder that, according to a poll conducted by the U.S. Chamber Technology Engagement Center

(C-TEC), 80% of Americans feel it's vital for the U.S. to lead the world in AI. The reality before us is as simple as it is stark: <u>whoever leads in the advancement of AI will lead the global economy</u>…"

Harvard Kennedy School BELFER CENTER published on Dec. 07, 2021 its landmark report "The Great Tech Rivalry: China vs the U.S.", authored by Graham Allison, Kevin Klyman, Karina Barbesino, and Hugo Yen. To quote:

"… In the longer-term competition, China's advantages begin with its population of 1.4 billion that creates an unparalleled pool of talent and data, the largest domestic market in the world, and universities that are graduating computer scientists in multiples of their American counterparts. China graduates four times as many bachelor's students with STEM degrees and is on track to graduate twice as many STEM PhDs by 2025. By contrast, the number of domestic-born AI PhDs in the US has not increased since 1990…

"Because a primary asset in applying AI is the quality of data, China has emerged as the Saudi Arabia of the twenty-first century's most valuable commodity. Even so, the United States enjoys two advantages in human capital that Beijing cannot replicate. First, half of the world's AI superstars work for U.S. companies. Second, America can recruit from all the world's 7.9 billion people, while inherent insularity restricts China to its own population…"

The Special Competitive Studies Project, set up in late 2021 by Eric Schmidt, former Google CEO, and Bob Work, former deputy defense secretary, reported September 2022:

"In the AI competition, the United States has a small lead with China catching up quickly across the AI stack…" 11.11.2022 18:06

In <u>State of AI Report</u> October 11. 2022, Nathan Benaich & Ian Hogarth reported: "The China-US AI research gap has continued to widen, with

Chinese institutions producing 4.5 times as many papers than American institutions since 2010, and significantly more than the US, India, UK, and Germany combined. Moreover, China is significantly leading in areas with implications for security and geopolitics, such as surveillance, autonomy, scene understanding, and object detection."

21.04.2023 19:41 p81/106 23250 words

(2) Quantum Science and Technology (S&T)

Published March 29, 2022 in **Neural/The Converation**, the article by Stuart Rollo, Postdoctoral Research Fellow, University of Sydney, has made various relevant and salient points which are reproduced as follows:

The quantum tech arms race is bringing us better AI and unhackable comms.

Quantum technology, which makes use of the surprising and often counterintuitive properties of the subatomic universe, is revolutionizing the way information is gathered, stored, shared, and analyzed.

The commercial and scientific potential of the quantum revolution is vast, but it is in national security that quantum technology is making the biggest waves. National governments are by far the heaviest investors in quantum research and development (R&D).

Quantum technology promises breakthroughs in weapons, communications, sensing, and computing technology that could change the world's balance of military power. The potential for strategic advantage has spurred a major increase in funding and research and development (R&D) in recent years.

The three key areas of quantum technology are computing, communications, and sensing (CCS). Particularly in the United States and China, all three are now seen as crucial parts of the struggle for economic and military supremacy.

Advances in quantum computing could result in a leap in artificial intelligence (AI) and machine learning (ML).

Superconducting quantum interference devices (or SQUIDS), which can make extremely sensitive measurements of magnetic fields, are expected to make it easier to detect submarines underwater in the near future.

At present, undetectable submarines armed with nuclear missiles are regarded as an essential deterrent against nuclear war because they could survive an attack on their home country and retaliate against the attacker. Networks of more advanced SQUIDS could make these submarines more detectable (and vulnerable) in the future, upsetting the balance of nuclear deterrence and the logic of mutually assured destruction (MAD).

HPC Wire reported October 12, 2022 "JP Morgan Chase (JPMC) Bets Big on Quantum Computing".

JPMC focuses on becoming quantum-ready soon.

JPMC has been busily developing quantum algorithms around optimization, machine learning (ML), natural language processing and publishing the results.

At a recent presentation by Marco Pistola, MD of JPMC's Global Technology Applied Research Center, the distinguished ex-IBM researcher addressed the theme Why financial services will perhaps be first to take advantage of quantum computing: "The reason is that in finance, we have a many use cases (that) have exponential complexity," said Pistola. ("The) level of complexity explodes as soon as a data set becomes big enough and a classical computer cannot solve that problem any more…"

Quantum advantage is expected by 2025.

Envisaged uses in financial services include risk analysis, derivative pricing (real time), targeting and prediction to provide efficient and enhanced credit scoring and accurate product recommendations, and portfolio optimization. And benefits, including real time results (seconds vs hours), greater accuracy more analyses performed, significantly more energy efficient.

According to the report by Jason Loh and Hazriq Iqmal in **THE EDGE MARKETS** October 19, 2022, a quantum computer could compute 158 million times faster than a supercomputer, or 4 minutes against 10,000 years.

While a 4,099-qubit quantum computer would take only 10 seconds to break a Rivest-Shamir-Adleman (RSA) encryption (one of the most secure encryption algorithms), a conventional computer would take 300 trillion years to do.

Where do they stand? To quote Graham Allison et al ("The Great Rivalry: China vs the U.S." Dec. 07. 2021): "In quantum information science (QIS), America has long been viewed as the leader, but China's national push presents a clear challenge. China has already surpassed the U.S. in quantum communication and has rapidly narrowed America's lead in quantum computing…"

In its November 4 2022 report on Recent Trends in Quantum Computing (QC), ONPASSIVE has stated in conclusion:

"Quantum computing has the potential to revolutionize industries such as finance, pharmaceuticals, AI, and automotive (autonomous transportation) in the next few years, fundamentally changing the world around us.

"The value of quantum computer (basically classical computer + quantum characteristics and some features related to quantum physics) comes from the quantifiable way they work. In the future, QC technology has the potential to transform many things, such as communications, cryptography, and computers..."

Dr Duncan Earl, President & Chief Technology Officer (CTO) of Qubitekk in Vista, California, has written in an article published in **physicsworld** 08 May 2023:"... The nation leading the race so far to harness the power of quantum technologies is China. The country has constructed a 2000km quantum-secured fibre-optic network and in 2017 demonstrated quantum-secured satellite communications. In 2019 and 2020 China's demonstration of "quantum advantage" -- a critical threshold on the path to powerful quantum computers -- further revealed a sophistication and acceleration of its quantum-information programme...

"The Chinese are indeed prioriitizing and dominating the quantum space, with every intention of winning this technology race..."

In the opinion piece titled "Why the US needs a 'quantum Oppenheimer' (a 'quantum tsar') to beat China in the quantum race", Earl contends the US can only win with a co-ordinated programme under a cohesive leadership.

Earl further stresses: "The US must urgently advance its national conversation from a techno-centric discussion to one focused on the timing and organizational challenges of accelerating progress for the nation's quantum programme." 10.05.2023 21:29 21:33

(3) Semiconductors

"Over the last 30 years, China's economic emergence has led to it challenging the United States for technological pre-eminence -- not unlike Japan in the 1960s to 1990s... By examining the semiconductor market, can we determine if China will succeed the U.S. in global technological leadership?," Sebastion Ko, MD of APAC, asked in guest blog post titled "The US-China Battle for Semiconductor Supremacy: A Case Study on Technology Competition", March 31, 2021 Fiscal Note Executive Institute.

"The semiconductor is a basic technological building block. No nation can claim technological leadership with(out) ample (adequate) access to its supply. Moreover, the challenges faced by the United States and China in the semiconductor industry mirror their technological race. The U.S. is maintaining the leadership position, while China is the top contender given the sheer scale of its commitment to technology advancement…"

Ko's conclusion: "The United States is about five to 10 years ahead in semiconductor R&D and manufacturing capability. While the U.S. and its allies are highly influential in the chip markets and are producing the majority of the talents in this area, they will likely adopt more strategies to slow down and contain Chinese technological progress.

"Lacking access to large talent pools and foreign partners with the necessary know-how, China should continue to lag behind the United States at least in the mid-term. But Chinese firms are resourceful and may catch up quickly; they are backed by the state and the Chinese government is committed to play the long game…"

BELFER CENTER's December 2021 Paper "The Great Tech Rivalry: China vs the U.S." has incorporated a progress report on China as well as America's response. To quote:

"With a three-fold increase in its share of global semiconductor consumption (from less than 20% in 2000 to 60% in 2019), China's growing domestic demand has provided both market and national security incentives to expand its push into the semiconductor industry, culminating in two notable successes. First, in semiconductor fabrication, China's share of global semiconductor manufacturing capacity has surpassed America's at 15%, up from less than 1% in 1990, while the U.S. share has fallen from 37% to 12%.

"The Semiconductor Industry Association (SIA) projects that over the next decade, China will develop 40% of new global capacity and become the world's largest semiconductor manufacturer, with 24% market share. Moreover, China's national champion in semiconductor fabrication, Semiconductor Manufacturing International Corporation (SMIC), has consistently ranked among the top five foundries over the past decade, and its breakthrough N+1 7-nanometer process last year (2020) means that its advanced fabrication capabilities now rival Intel's.

"Second, in the chip design arena, Huawei's HiSilicon subsidiary has grown into an integrated circuit (IC) design powerhouse. In 2020, it became the first Chinese company to break into the top ten semiconductor companies and replaced long-time market leader Qualcomm as China's top smartphone processor supplier, though export controls have damaged the company's near-term prospects.

"While China is still dependent on semiconductor imports to meet 85% of domestic demand, these recent achievements disprove the decades-long conventional wisdom that China's semiconductor industry cannot catch up. Indeed, by (TSMC Founder Morris) Chang's best judgment, China is only "one to two years behind the U.S. and Taiwan" in chip design and "five years behind TSMC in fabrication..." P69 15059 words 07.11.2022 20:03

To quote BELFER CENTER's report on the US:

"With 48% overall industry market share compared to China's 5%, the U.S. is the undisputed global leader in semiconductors. Yet, its positions in the design and fabrication arenas have weakened significantly ... only 44% of U.S. -designed chips are fabricated domestically today. Both Intel and GlobalFoundries are far behind in the next-generation chip competition, leaving 90% of advanced fabrication in the hands of Taiwan-based TSMC…

"In semiconductor manufacturing inputs, America retains a strong position and controls key supply chain choke points through firms like Applied Materials and Lam Research, holding 55% market share of semiconductor manufacturing equipment (versus 2% for China) and 85% of electronic design automation (EDA) software.

"Together, these trends suggest that while the U.S. will not be displaced as an industry leader in the near term, China has made strong progress in two of three critical areas that, if maintained, could see China's semiconductor industry grow rapidly within the next decade. Though recent U.S. actions like sanctions on Huawei and SMIC's inclusion in the Entity List have slowed China's progress, completely cutting off China's access to advanced semiconductors would be a self-sabotaging policy since the Chinese market accounts for 36% of all U.S. chip sales."

On China's measured response to US technology and trade restrictions, BELFER CENTER has reported: "Since 2019, the Commerce Department has added over 300 Chinese companies including Huawei and SMIC to the Entity List, a trade blacklist which restricts access to national security-controlled goods and technologies. Most notably, the Entity List designation cut off Huawei's purchases of advanced semiconductors that were made using U.S.-sourced equipment and SMIC's access to U.S.-made semiconductor manufacturing equipment for leading-edge chips.

"Due to these restrictions, Chinese firms have been researching replacements to American technology and achieving self-sufficiency (or self-reliance) where possible. For example, Huawei's HiSilicon has developed its first chipset based on the open-source RISC-Y architecture as an alternative to the more commonly used ARM architecture, which arguably falls under the Entity List restrictions…"

As reported by Canadian tech media outlet **Tech Insights**, SMIC appears to have used 7nm technology to make the Mines Va Bitcoin Miner system on chip (SoC). In December 2020 SMIC was put in the Entity List by the US Bureau of Industry and Security (BIS) to block it from the necessary Extreme Ultraviolet (EUV) Lithography machine made in the Netherlands and to prevent China's largest chipmaker from stepping to advanced modes of 10nm and below.

According to **Tech Insights**, SMIC products made from the quasi-7nm process had been shipped for a year. (Che-Jen Wang August 20, 2022 **THE DIPLOMAT**) In October, it was reported that SMIC had successfully developed "quasi-7nm" chips with the FinFET N+1 process, using Deep Ultraviolet (DUV) Lithography machine.

Of the total $186.5 billion worth of semiconductor chips consumed in China in 2021, only 16.7% were produced locally by both foreign (10.9%) and domestic (6.6%) companies. China's self-sufficiency came to only 16.7%.

"… Chinese chipmaker SMIC, which recently shocked the US by announcing that it had produced 7nm chips despite being denied access to EUV equipment, is now reported to be advancing to more advanced 5nm. SMIC has also started construction of a new 300 mm (12-inch) wafer fab," Scott Foster reported October 5, 2022 **ASIA TIMES**.

"In the fields of artificial intelligence (AI) and high-performance computing (HPC), Xiangdixian Computing Technology and Moffett AI have announced new devices that they claim can replace the GPUs that Nvidia and AMD are no longer allowed to sell to China. The design rules are not as advanced (12nm versus 4nm for Nvidia) but they work.

"Nikon's NSR-S635E ArF immersion scanner "Provides world-class device patterning and productivity for 5nm node applications and beyond," the company has said, revealing new chip-making horizons without EUV.

"If successful, US efforts to stop ASML from shipping DUV lithography tools to China might be the best thing that ever happened to SMEE (Chinese lithography equipment maker, reportedly working on a new ArF immersion lithography tool used to make 7nm chips).

"Industry association SEMI lists about 80 Chinese companies involved in semiconductor equipment research and manufacturing -- and all of them can be assumed to receive government support.

"Following in the footsteps of Japan, but driven by fear of escalating sanctions, China now aims to develop a complete autonomous semiconductor supply chain."

According to **IC Insights**, China consumed US$186.5 billion worth of semiconductors in 2021, accounting for 35.5% of the world market. Only 17% of Chinese semiconductor demand was met by production in China and only 7% by Chinese companies.

"These figures show the market opportunity for Chinese semiconductor design, manufacturing, and production equipment companies, and the corresponding opportunity cost for foreign companies hamstrung by US government export restrictions. Import substitution alone can give Chinese companies economics of scale."

In an article headlined "Washington Raises Stakes in War on Chinese Technology" following US Commerce Department's highly calculated campaign to deny advanced chips and critical technologies to China, Edward Alden wrote October 18, 2022 (COUNCIL ON FOREIGN RELATIONS): "… With China so integral to the global electronics supply chain -- and to the profits of Western technology companies -- Washington has been trying to find a balance between treating Beijing as an economic partner and a geopolitical leader. But a choice has now been made: For the first time in a generation, weakening China is now more important to the United States than working with China..."

Alden quoted Boston Consulting Group (BCG), which has estimated that a complete ban on US chip sales to China would cost American semiconductor firms 18% of their global market share and 37% of their revenue.

"Take semiconductors: Nothing is more "core" than the integrated circuits (ICs) that are the brains of all electronics, from personal computer CPU, to navigation systems for ballistic and cruise missiles," American Enterprise Institute (AEI) commented October 20, 2022.

"In 2021, microelectronics were China's top import, even above oil -- and this is not for lack of trying to meet demand indigenously... To be sure, it is improving in complex manufacturing (fiercely competing with Taiwan Semiconductor Manufacturing Company) and artificial intelligence (AI) chip design (competing with American players). Still, China is far behind, and will be an importer for some time to come..."

Postscript (PS)

"... Washington has been nudging East Asian partners, Taiwan, South Korea, and Japan to form a "Chip 4" industry alliance to isolate China from the international tech ecosystem, and bolstered efforts to develop its homegrown industry with the passage of the CHIPS Act (August 2022)," reported Liam Gibson, a Taipei-based analyst, columnist and journalist. (9 Sep 2022 **ALJAZEERA**)

"Semiconductors have emerged as one of the fiercest battlegrounds in the intense rivalry between the US and China. Beyond functioning as the lifeblood of the modern economy, powering everything from iPhones to fighter jets, the chips are seen as critical to unlocking the technological breakthroughs of the future, meaning tomorrow's global balance of power would rest on the wafer-thin (wafer-based) chips being developed today..."

Chris Miller, assistant professor of international history at The Fletcher School of Law and Diplomacy, and author of "Chip War: The Fight for the World's Most Critical Technology" (originally published October 4, 2022), told **Al Jazeera**: "The US is trying to reinforce its central role in the world's semiconductor ecosystem and ensure that China is unable to produce the most cutting edge chips... Control over semiconductors will not only shape the future of the world economy, from cloud computing to autonomous driving; they are also fundamental to military power..."

Wu Jincheng, director of the Shanghai Municipal Commission of Economy and Digitalization, told a press conference September 14, 2022, that Shanghai-based semiconductor firms have achieved mass production of 14-nm chips and made breakthroughs in 90-nm lithography machines, 5-nm etching machines, 12-inch (300 mm) silicon wafers, central processing units (CPUs), and 5G chips.

According to Chen Jia, an independent research fellow on strategy, large-scale production of 14-nm chips will contribute considerably to the development of new-energy vehicles (NEVs), smart cities, intelligent manufacturing, and the Internet of Things (IoT).

According to Robert Castellano, author of <u>Semiconductor Deep Dive</u>, SMIC has been developing the 5-nm node chips without embargoed EUV lithography since 2021 and will reach production of 5nm in 2025. (**Seeking Alpha** Sep. 15, 2022)

"China's stated goal is to become self-sufficient in the production of semiconductors for its domestic market and to develop technology that is competitive on the world market," Castellano commented.

"The US had a chance to stop China from being a technology powerhouse and ultimately an existential threat 20 years ago but buckled under lobbying from industry consortium SEMI to benefit its semiconductor equipment members…"

Speaking at a meeting of the Foreign and National Defence Committee on 12 October 2022, Taiwan's National Security Bureau Director General Chen Mingtung denied that current defence plans included the destruction of TSMC's plants or evacuation of engineers. (**Army Technology** Features November 30, 2022)

Chen described TSMC's global supply chain as reliant on industrial partners in other countries including the Netherlands, and said that the US does not need to destroy TSMC's factories because severing supply chains for vital components would be enough to halt production.

Speaking at the Richard Nixon Foundation Grand Strategy Summit on 10 November 2022, former US National Security Advisor Ambassador Robert O'Brien said "I don't think we would even allow" China taking

Taiwan and its chip factories "intact" to have a monopoly over "advanced computer chips"
(accounting for 95% of US semiconductor consumption).

P74 16108 words 08 November (Ma's day) 2022 14:0008.02.2023 17:51
P93/106 23293 words 24.04.2023 22:42

Appendix (A)

China's Chips Challenge

In mid-June 2021, President Xi Jinping appointed his top economic advisor (childhood friend and right-hand man) Vice-Premier Liu He to realize the country's goal of achieving self-reliance in technology as well as to spearhead the development of third-generation chips for self-sufficiency in the vital semiconductor industry. (**Verdict.co,uk** June 17, 2021)

Vice-Premier Lui He has reportedly described the IC industry as "the core nexus of modern industrial systems", of considerable significance to "national security and the progress of Chinese-style modernization". (**Xinhua/South China Morning Post** Mar 03, 2023)

According to Liu He, China has already developed "a relatively complete semiconductor supply chain" with "very strong capabilities at home in some segments" and with the huge domestic market as "the most precious resource for China and a strategic advantage to promote the semiconductor industry".

The twin goals are complete semiconductor self-sufficiency and technological self-reliance by 2035.

"China aspires to lead the world in developing the chip technology that will power platforms, such as those for 5G and artificial intelligence (AI), that are central to the country's economic and high-tech goals," Caroline Gabriel has observed and written 06 May 2021. Gabriel is Research Director in Analysys Mason, a world-leading UK-based research and management consultancy focused on telecoms, media & technology (TMT).

"These two goals, for greater self-reliance and for technology leadership, are driving a significant increase in investment and innovation, which will over time, transform China's hi-tech sector and affect the global ecosystem," Gabriel has reported.

"Progress is not only driven by the current geopolitical situation, but also by broader ambitions to be at the cutting edge of technologies that will transform society and the economy, such as 5G, AI, robotics and green energy.

"In an assessment carried out by Analysys Mason, memory technology and AI platforms were judged to be the areas where Chinese companies had made the most significant progress, by 2020, in closing the gap with international suppliers in terms of technology and production scale...

"In some areas of chip development, China is close to equalling the international market. It is ahead of the world in transcoding and crypto-mining; and in some broader categories, including some types of memory technology, it is levelling with the global giants.

"However, the challenge for China's ambitions lies not in a lack of hi-tech capabilities, but in the time it will take to scale up its innovations and processes to support huge levels of demand. Chinese companies purchased chips worth UDS380 billion in 2019, so the goal of self-sufficiency is aq challenging one...

"In reality, China does not need to achieve self-sufficiency and technology parity across the whole vast range of semiconductors. The biggest impact on its economy will be felt if it becomes innovation on chips that power the most critical advanced technologies..."

The world=leading consultancy has cited memory, 5G SoC (system-on-chip) and AI chips as well as cloud processors and accelerators, and ultra-low power chips for mass-scale IoT uses.

Time to 75% self-sufficiency and time to technical leadership vary from 1 year to 4 years.

The experts have concluded that "the maximum impact will be felt with a return to <u>international cooperation</u>, enabling the hi-tech industry worldwide <u>to take advantage of China's innovations</u>, and allowing Chinese companies to access markets and hi-tech developments on a global basis."

In brief, a timely and win-win return to <u>global normal</u>!

P96/109 24253 words 26.04.2023 21:17

As reported by Lara Williams 25 July 2022 **Investment Monitor**, China has expertise in developing new super conductive materials. Chinese telecommunications giant Huawei is leding in photonic computing as well as in development of transistors made of graphene.

China also leads in outsourced semiconductor assembly, testing and packaging (ATP). The development of advanced chip packaging, new transistor architecture and new carbon-based materials could be game changers to make China the global leader in the semiconductor industry by 2025.

The majority of global demand will reportedly continue to be for 28-nm .chips and above, rather than the more advanced 10-nm chips or below. And, China is becoming more self-sufficient as domestic manufacturers including SMIC, Hua Hong and ASMC increase foundry capacity with at least seven major new fabs coming on stream by 2024.

While no country can feasibly create a completely self-sufficient domestic semiconductor supply chain within 15 years, GlobalData has predicted that China will become the world's leading semiconductor superpower by 2030 through a combination of its growing market and production capacity.

Appendix (B)

CHIPS for America

On 25 April 2023, CHIPS Research and Development Office published its 30-page paper on CHIPS for America Vision and Strategy in semiconductor R&D and in establishing the National Semiconductor Technology Center (NISTC) -- to further promote and perpetuate US leadership in this vital industry.

To quote: "It is hard to imagine a sector more important to the national and economic security of our nation than the semiconductor industry. Nearly every innovation of the future, whether for smarter transportation or better medical devices, for agricultural efficiencies or to address climate risks, or for the technology that powers our national defense capabilities, will depend on semiconductor technology.

:The Biden adminstration called out three classes of technology as being particularly critical over the coming decade: computing-related technologies including microelectronics, quantum, and artificial intelligence (AI); biotechnologies and biomanufacturing, and clean energy technologies. Each of these fields is dependent on semiconductors, and semiconductor advancements, in turn, are driven by the demands of new applications.

"The pace of innovation in the semiconductor technology sector over the past seven decades has been extraordinary. The industry has progressed from building a few transistors in silicon to, today building billions of transistors on a single wafer. Much of this success is due to advances in the manufacture of ever smaller semiconductors following a consistent progression in scaling known as Moore's Law.

"Today, the smallest dimensions of leading-edge semiconductor devices have reached the atomic scale and the complexity of the circuit architecture is increasing exponentially with the use of three-dimensional structures, the incorporation of new materials, and improvements in the thousands of process steps needed to make advanced chips.

"Into the future, as new applications demand higher performance semiconductors, their design and production will become even more complex. This complexity makes it increasingly difficult and costly to implement innovations because of the dependencies between design and manufacturing, between manufacturing steps, and between front-end and back-end processes.

"To accelerate innovation in semiconductor technology, there is a need for a systems-level research and development (R&D) approach that connects sophisticated tools, resources, and facilities. This provides innovators with the flexibility to explore improvements to complex heterogeneous systems and the confidence that new designs and manufacturing technologies will be successful.

"More than ever, research and development (R&D) activities need to be closely aligned with design and manufacturing processes...

"Recognizing these needs, Congress appropriated funds for a national semiconductor technology center (NSTC) to support and extend U.S. leadership in semiconductor research, design, engineering, and advanced manufacturing. By integrating efforts across a complex ecosystem, a successful NSTC will advance critical semiconductor research and development (R&D), expand access to design and manufacturing resources and allow industry, academics, and government to build on each other's work (synergistically), abd reduce the time and cost of bringing technologies to market.

"As an independent public-private consortium, the NSTC will provide a (joint) platform where government, national laboratories, industry, customers, suppliers, educational institutions, entrepreneurs, workforce representatives, and investors collaborate…."

The NSTC Mission: As per CHIPS Act of 2022, "The secretary of Commerce, in collaboration with the Secretary of Defense, shall establish a national semiconductor technology center (NSTC) to conduct (world-class) research and prototyping of advanced semiconductor technology and grow the domestic semiconductor workforce to strengthen the economic competitiveness and security of the domestic supply chain. Such center shall be operated as a public private-sector consortium with participation from the private sector, the Department of Energy, and the National Science Foundation…"

The National Institute of Standards and Technology (NIST) of the US Department of Commerce has described CHIPS for America semiconductor R&D strategy as part of President Biden;s investing in America Agenda to advance US competitiveness and technological leadership.

Secretary of Commerce Gina Raimondo has said: "While the manufacturing activities of the CHIPS Act will bring semiconductor m anufacturing back to the U.S., a robust R&D ecosystem led by the NSTC will keep it here (at home)…"

P100/113 25079 words 27.04.2023 22:33

Addendum

Strategic Shift
From Collaboration to Competition

(A) The 1999 Cox Report's "Dirty Little Secret"

"... From the very outset of the Sino-U.S. relations in the early 1970s, successive Republican and Democratic administrations believed that the enhancement of Chinese power -- as a counterbalance to Soviet power -- was in the national security interest of the United States, and persistently sought to advance this goal in the ensuing two decades. This U.S. commitment was repeatedly imparted to senior Chinese officials in both word and deed. The United States was intent on strengthening relations in a host of highly sensitive areas and shaping China's expectations of the United States and its pursuit of U.S. high technology (HT). The Chinese may well have exploited these opportunities by all available means, but they were walking through a door that the U.S. government had long since decided to open," Jonathan D. Pollack, a senior advisor for international policy at the Rand Corporation think tank in Santa Monica, California, has written in a concise and insightful review of the controversial Cox Report on Chinese espionage and illicit technology acquisition. (Arms Control Association, April/May 1999) The Cox Report did not disclose bilateral active intelligence collaboration and American transfer of technology and its contribution to the skill base of the Chinese aviation industry.

"From the earliest years of Sino-American relations, senior U.S. officials (including their National Security Advisor Henry Kissinger) provided Chinese interlocutors with highly sensitive U.S. intelligence data on Soviet military capabilities and deployments -- without the Chinese having ever solicited this information," Pollack narrated.

"The Nixon administration (1968-1974) sought to consolidate its diplomatic and security actions with China in Vietnam, in South Asia, and in relation to Taiwan, Japan and Korea. Under President Ford (1974-1977), the United States explicitly encouraged major European allies to relax their export policies toward China, including on weapons sales; the Chinese were regularly kept informed of these actions. U.S. policymakers also sought to find the means to permit sales of U.S. computers to China that it would not export to the Soviet Union...

"The report also fails to acknowledge that the Reagan administration (1981-1989) initiated foreign military sales (FMS) programs to China, including sales to military end-users, provision of technical know-how to Chinese military research and development (R&D), and active collaboration between U.S. defense contractors and Chinese counterparts. By 1987, these programs comprised avionics packages for Chinese combat aircraft, sales of anti-submarine warfare torpedoes and gas turbine engines for the Chinese navy (the latter still in use on Chinese destroyers), sales of artillery-locating radar and the upgrading of artillery production capabilities...

"Additional developments throughout the latter half of the 1980s -- including purchases of more advanced computers, acquisition of sophisticated machine tools, prospective sales of nuclear reactors, and an explosive growth in the training of Chinese scientists and students in the United States -- attested to the consolidation of U.S.-China relations, with the ever-increasing focus on advancing China's technical and industrial capabilities, many with potential relevance to China's military modernization..."

Subsequently, however, the 1989 Tiananmen episode "shattered the working consensus between the executive branch and Congress (in Washington) on closer U.S.-China relations."

To further quote PollacK: "Equally significant, the collapse of the Soviet Union eroded the strategic assumptions underlying U.S.-Chinese defense collaboration. The Bush administration (H.W. 1988-1993) sought to retain some of these dealings, but changes in U.S. strategy were already underway, including a substantial augmentation of U.S. military sales to Taiwan as the Chinese increasingly pursued weapons purchases from Russia. The Clinton administration (1993-2001), with intermittent success but also some major setbacks, sought to restore a modicum of civility in Sino-U.S. military relations but it was not prepared to revisit the post-Tiananmen sanctions precluding military sales to China. At the same time, the scope of U.S. defense links with Taiwan continued to grow during the 1990s.

"Thus, the Cox Report appears when the strategic divergence between the United States and China has widened, and when the Chinese are increasingly assertive about longer-term modernization goals that are no longer animated by animosities with Moscow. Indeed, Russia and China share common cause in seeking to inhibit, or at least caution, the United States from the unilateral exercise of its military power. One does not need to subscribe to the more conspiratorial Chinese renderings of U.S. strategy to discern the essential Chinese logic: officials in Beijing see the United States as intent on impeding the growth of Chinese national power and the enhancement of its strategic position..."

(B) <u>2003 Report by the Council on Foreign Relations</u>
    Task Force o<u>n China's Military Modernization</u>

Chaired by Dr Harold Brown, former US Secretary of Defense (1977-1981), the well-resourced Task Force has principally addressed the impact of China's military modernization on: (1) the military balance between China and the US, (2) the US commitment to the defense of Taiwan, and (3) the Chinese nuclear deterrent vis-a-vis the US military superiority ("in the eyes of the beholder" to quote Michael Pillsbury, director of the Center on Chinese Strategy at the Hudson Institute).

(1)   "The Council on Foreign Relations Independent Task Force on Chinese Military Power finds that the People's Republic of China

(PRC) is pursuing a deliberate and focused course of military modernization but that it is <u>at least two decades behind</u> the United States in terms of military technology and capability. Moreover, if the United States continues to dedicate significant resources to improving its military forces, as expected, the balance between the United States and China, both globally and in Asia, is likely to remain decisively in America's favor beyond the next twenty years (2023)..." -- Report in <u>Chinese Military Power</u> 6/12/03)

(2)   "The PLA currently has the ability to undertake intensive, short-duration air, missile, and naval attacks on Taiwan, as well as more prolonged air and naval attacks. The efficiency of either scenario would be highly dependent on Taiwan's political and military response, and especially on any actions taken by the United States and Japan," the CFR Task Force reported (p,3).

"Although the U.S. forces would ultimately prevail in a military crisis or conflict, Beijing might be able to impose serious risks and costs on the U.S. military if the United States concluded that it was necessary to commit air and naval forces to battle with China in defense of Taiwan..." (p. 4)

The Task Force then reported: "Any conflict across the Taiwan Strait would have extremely adverse impact on the strategic landscape in Asia, regardless of the outcome. Therefore, the most critical aim of U.S. strategy in the cross-strait situation must be to deter and minimize the chances that such a crisis will occur.

"Taiwan is fundamentally a <u>political</u> issue, and any effective strategy must coordinate military measures designed to deter with diplomatic efforts so as to reassure both China and Taiwan credibly that their worst fears will not materialize. For U.S. policy toward Taiwan, this means providing Taiwan with the weapons and assistance deemed necessary for the creation of a robust defense capability and not making a deal with Beijing behind Taipei's back.

"For U.S. policy toward China, this means maintaining the clear ability and willingness to counter any application of military force against Taiwan while conveying to Beijing a credible U.S. commitment to not support Taiwan's taking unilateral steps toward de jure independence." (p. 4)

(3)   On nuclear deterrence: "The Task Force expects that the United States will continue to possess overwhelming dominance over China's nuclear forces for the foreseeable future. China, however, is improving the survivability of its small retaliatory "countervalue" deterrent force (targeted at aggregated urban population and essential infrastructures).

"China's nuclear arsenal will likely expand in number and sophistication over the next ten to twenty years.

"Although the Task Force is uncertain about the specific impact of U.S. missile defense plans on Chinese nuclear modernization in terms of numbers and force deployment, we believe that China will do whatever it can to ensure that a U.S. missile system cannot negate its ability to launch and deliver a retaliatory second strike…" (p.

(C) US Department of Defense (DoD) 23 July 2003
    Annual Report to Congress THE MILITARY POWER
    OF THE PEOPLE'S REPUBLIC OF CHINA
    addressing current and future military strategy
    and probable future course of military-technological
    development through the next twenty years.

Extracts from the 2003 DoD Report:

(1)   Beijing states that the current number one strategic priority is economic development. In addition to the important function it plays in raising living standards, economic development is regarded

as an important step in gradually increasing China's international leverage and military modernization.

An economically strong China also, over time, would enhance its relative CNP (comprehensive national power) and could allocate its resources for a more favorable "strategic configuration of power" (or "shi", to preserve national independence and help build "momentum" in increasing national power)... (p. 10)

(2)   While stressing the primacy of economic power, Beijing views the military as necessary to ensure that China's economic power will rise, to protect national interests, and to support China's eventual emergence as a great power and the preeminent power in Asia. (p. 14)

At the 16th Party Congress in 2002, China's leaders reaffirmed their primary commitment to economic development as well as their continued support for military modernization.

(3)   China's leaders believe that national unity and stability are critical if China is to survive and develop as a nation. China's leaders also believe they must contain conditions of state sovereignty and territorial integrity. (p. 51)

(4)   While seeing opportunity and benefit in interactions with the United States -- primarily in terms of trade and technology -- Beijing apparently believes that the United States poses a significant long-term challenge. (p. 51)

(5)   In support of its overall national security objectives, China has embarked upon a force modernization program to diversify its options for use of force against potential targets such as Taiwan, the South China Sea and border defense, and to complicate United States intervention in a Taiwan Strait conflict. (p. 51)

(6) Preparing for a potential conflict in the Taiwan Strait is the primary driver for China's military modernization. While it professes a preference for resolving the Taiwan issue peacefully, Beijing is also seeking credible military options. Should China use force against Taiwan, its primary goal likely would be to compel a quick negotiated solution in terms favorable to Beijing. ((p. 51)

(7) China is developing advanced information technology (IT) and long-range precision strike capabilities, and looking for ways to target and exploit the perceived weaknesses of technologically superior adversaries (following Serbian exploits in the 1999 Operation ALLIED FORCE in the Kosovo conflict in Yugoslavia).

(8) Beijing has greatly expanded its arsenal of increasingly accurate and lethal ballistic missiles and long-range strike aircraft that are ready for immediate application should the PLA (People's Liberation Army) be called upon to conduct war before its modernization aspirations are fully realized.

(9) China's force modernization program is heavily reliant upon assistance from Russia and other states of the Former Soviet Union (following its collapse and dissolution in 1991).

(10) (Back in 1981, China's deployment of its first ICBM (intercontinental ballistic missile) with a 4-5 megaton warhead and a striking range of 13,000km marked its emergence with an incipient nuclear deterrent force. By Pentagon's admission, the "third rate power" had earlier "nearly pushed off" the US forces from the Korean peninsula in the 1950-1953 Korean War.)

(11) In 2003, China operated about 126 ballistic missiles including estimated 20 DF5/5A ICBMs, capable of striking the continental United States (Robert S. Norris and Hans M. Kristensen of the Natural Resources Defense Council (NRDC), **Bulletin of the Atomic Scientists**, November/December 2993).

"China has had the technical capability to develop multiple reentry vehicle systems (MRVS) for 20 years (since early 1980s), including a multiple independently targetable reentry vehicle (MIRV) system, but has chosen not to do so," Norris 20:Kristensen reported. (China started to MIRV in 2015.)

P93 21035 words 20.02.2023 20:22

(D) US Department of Defense (DoD)
2022 Annual Report on China's Military Power

Main points:

(1)   The 2022 National Security Strategy (NSS) identifies the People's Republic of China (PRC) as the only competitor with the intent and, increasingly, the capacity to reshape the international order -- the PRC presents the most consequential and systemic challenge to U.S. national security and the free and open international system. (p. 1)

(2)   This report illustrates how the CCP (Chinese Communist Party) increasingly turns to the PLA (People's Liberation Army) in support of its global ambitions, and the importance of meeting the pacing challenge presented by the PRC's increasingly capable military. (p. 2)

(3)   The PRC increasingly views the United States as deploying a whole-of-government effort meant to contain the PRC's rise, which presents obstacles to its national strategy. (p. III)

(4)   PRC leaders believe that structural changes in the international system and an increasingly confrontational United States are the root causes of intensifying strategic competition between the PRC and the United States. (p. III)

(5)   The PLA seeks to modernize its capabilities and improve its proficiencies across all warfare domains so that as a joint force, it can conduct the full range of land, air, and maritime, as well as nuclear, space, counterspace, electronic warfare (EW), and cyberspace operations. (p. VI)

(6)   In 2020, the PLA Air Force (PLAAF) operationally fielded the H-6N strategic bomber, providing a platform for the air component of the PRC's nascent nuclear triad. In addition, the PLAAF is seeking to extend its power projection capability with the development of a new H-20 stealth strategic bomber. (p. 60)

According to Admiral Sam Faparo, head of US Pacific Fleet, China has equipped its 6 Jin-class Type 094 nuclear missile submarines (SSBNs) with the new JL-3 intercontinental submarine-launched ballistic missiles (SLBMs) with a range of over 10,000km. (Liu Xuanzun Nov 20, 2020 **GLOBAL TIMES**).

(7)   The PLA is developing intercontinental ballistic missiles (ICBMs) that will significantly improve its nuclear capable missile forces with more survivable systems. The new DF-5B and DF-41 ICBMs can each carry up to five nuclear warheads. (p. 65)

(8)   The PRC's ICBM arsenal consists of approximately 300 ICBMs, including fixed and mobile launchers capable of launching unitary and multiple reentry vehicles. (p. 65)

According to data compiled by Matt Korda and Hans M. Kristensen (**Bulletin of the Atomic Scientists** November 15, 2021), China has deployed over 200 nuclear warheads capable of striking the continental United States.

According to 2021 Report by US Department of Defense (DoD), the PLA is accelerating expansion of its nuclear arsenal of up to 700 warheads by 2027 and at least 1,000 by 2030.

(9)　On 27 July 2021, China conducted the first fractional orbital launch of an ICBM with a hypersonic glide vehicle (HGV) which flew over the world and returned inside China after flying over 40,00km and over 100 minutes. According to US military officials, the HGV did not hit its target, but came close. (p. 65)

(10)　The PLA is pursuing next-generation combat capabilities based on its vision of future conflict, which it calls "intelligentized warfare", defined by the expanded use of artificial intelligence (AI) and other advanced technologies at every level of warfare.

P96/99 21565 words 21.02.2023 21:57
P103/106 23244 words 21.04.2023 20:14

www.ingramcontent.com/pod-product-compliance
Lightning Source LLC
Chambersburg PA
CBHW021442210526
45463CB00002B/615